Automatic Patent Analysis - Examples

APA

Henri Dou

University Professor – International Consultant

January 2014

Editor CIWORLDWIDE

[Tapez un texte]

[Tapez un texte]

[Tapez un texte]

1 About this book

The development of the competition between enterprises, institutions and other entities around the world, brings the concept of innovation as one of the main way to create or maintain competitive advantages. In another hand, the development of the concept of frugal innovation[1] (or innovation Jugaad) will push to the search of information able to help the simplification and the creation of tailor made products robust enough to satisfy the need of the emergent market customers. The search for relevant information in this domain, underlined the role of the patent information which will help not only to boost innovation but also to described some idea of improvement in various domains (mainly mechanic, electronic, building, ..). The role of "utility models" which already count for half of the patent Chinese production is a good example of Jugaad information retrieved from patents.

But, the number of patent increases every years, and even if not all the firms patent their innovation, the marge number of inventions protected and then described in patent is large enough to help people in their search for innovation. To do so, the classical way to search for the best relevant patents (that is to say for the smaller but relevant number of patents) will not apply. To get the best possible information on a technological area, or type of products, etc. it is necessary to do it in another way. The APA (Automatic Patent Analysis) will provide the solution[2]:

- Perform a search large enough to encompass all the whereabouts of the subject. The result will be generally a large amount of patents.
- Upload these patents into your computer (you must use a system enable to upload the data fast enough
- Perform on the local database that you created the APA by combining all the data available in a patent description

---

[1] L'intelligence économique à l'heure du Jugaad, Henri Dou, Amazon format Kinfdle, http://www.amazon.fr/Lintelligence-%C3%A9conomique-%C3%A0-lheure-Jugaad-ebook/dp/B00HR5Q5Y6

[2] Automatic Patent Analysis (APA) to Improve Innovation and Decision Making in Science and Technology, Dou H., J. Kister, B. Mannina, International Journal of Latest Research in Science And Technology, Volume1, issue4, 2012

- Out of these combinations (list matrix, etc.) evaluated the impact of the most relevant patents (or companies or inventors or applications) on the strategy of your company (or laboratory, or regional development, etc.)

With most of 90 million patents, the world patent database, available from the EPO (European Patent Office), or the US patent available from the USPTO (Us Patents and Trademark Office), will enable you to get the best overview of the technologies developed in various fields. Moreover, if the classical patent bibliographic data is available, it is also possible to upload the full text of the patent (is present in the EPO or USPTO databases) and to look into the register (from the EPO) or the Inpadoc status of the patents retrieved.

Unfortunately, in academic works, the patents are seldom cited and used as an information source in the academic laboratory. The SMEs also do not use enough the patent information even if the data are freely available. The purpose of this book is to show through various examples, that APA is an interesting area of information management, and that with the facility available (for instance Matheo-Patent from the Matheo-Software company[3]), the development of Automatic Patent Analysis is easy to used. Most of the examples described have been presented in various meetings and never published up to now. They can be read independently one from the other and the subjects show that this technique may applied to almost all the fields.

In some analysis the geographical mapping of the patent countries involved in the analysis are presented. When you see these mappings you must be aware that for WO and EP patents, this is a potential representation since for these patents when they are granted, the patent applicants is returned to the national patent offices. Then he must select the countries that he wants to protect to fulfill the national administrative conditions to get the protection.

We recommended also to the readers, to make a good usage of the Chinese patents[4] (China is the most producing countries in the patent field) and to examined carefully

---

[3] http://www.matheo-software.com a demo version is available on this site

[4] Chinese Patent - A Tentative Explanation of Various Strategies of Patenting
Dou Henri, Dou Jean-Marie Jr

the "utility models", which are small inventions with a span of 10 years, and most of the time not extended in other countries (as patents). They may provide you with very good ideas.

For the users which will be interested in using patent information for regional or industrial development, more information is available on the site http://www.ciworldwide.org as well as in the following electronic books. [5] [6]

The choice, for publication of the ebook kindle format of Amazon has been done to facilitate the link between all the references present in the book and their development of pdf access using the url links. The choice of the ebook format has also been done, because the price of the book is suitable for most of the users.

---

[5] Competitive Intelligence and Regional Development. A focus on Developing countries, Henri Dou, Jean-Marie Dou Jr.Sri Damayanty Manullang, Amazon format Kindle, 2012 http://www.amazon.fr/Competitive-Intelligence-Regional-Development-Developing-ebook/dp/B00AWBIBWQ/ref=sr_1_5?s=digital-text&ie=UTF8&qid=1390137216&sr=1-5

[6] Competitive Technical Intelligence - A Focus on Industry Development in Developing Countries, Sri Damayanty Manullang, Jean-Marie Dou Jr. and Henri Dou, Amazon format Kindle, 2013, http://www.amazon.fr/Competitive-Technical-Intelligence-Development-Developing-ebook/dp/B00B24QMC8/ref=sr_1_6?s=digital-text&ie=UTF8&qid=1390137216&sr=1-6

2 Implementing strategic information by using APA (Automatic Patent Analysis) - Example in the field of corrosion resistant steel

Henri Dou*, Daniel Coelho**,

*Director of Atelis (Strategic Work Room of the ESCEM Group, France),email
douhenri@yahoo.fr

** PhD Student, Brazil, Daniel@coelho.org

1 - Introduction

This paper is done to show how patent analysis is a useful information tool to determine the environment of a research subject before the beginning of laboratory experiments and measures. Patents are now available freely and there are several patent databases available from Internet. Different tools are available to query the database, download all the bibliographic data and perform off line an automatic patent analysis (APA). We choose to work for this purpose with the Matheo-Patent software available from the company Matheo-software (1). The data base which will be used will be the EPO WorldPatent Database which covers more than 80 countries. The patents are particularly interesting since they are almost the only documents which build a bridge between the fundamental research and its applications. Moreover, what is published in patent documents is seldom published elsewhere. Another particularity of the patent documents is that most of the time they are not consulted in fundamental research, a simple analysis of the bibliographies in research papers second this point of view.

2.1 Material and Methods

2.1.1 Materials

As we saw above the database which will be used is the worlwide database (2) from the EPO (3)(European Patent Office). The query of this database can be done by using words from titles and abstracts (there is no keywords in patents references), the Applicant names, the inventors names, the patent dates, the priority patents, the patent numbers and the IPC (International Patent Classification). When a local database is created from the patents downloaded online, this database is formatted to be able to be analyzed automatically (4, 5). In the same time the significant words from the titles and abstracts are extracted as well as the drawings available (the first page drawings). When this is done, all the bibliometrics (6) analysis can be done automatically lists, charts, matrices, networks. These different treatments will provide an implicit information, generally new, that it is impossible to detect by reading or browsing the database. These treatments allows to create groups of patents and further analyzed their content, to select significant patents, to benchmark automatically the knowhow of companies and inventors, to determines the inventors networks of the company's networks, etc. The local database created can be updated using the same query or completed by different queries, the patents will be added to the initial database and the duplicate eliminated. If the user wants to work with a large database, selections (groups) according words, dates, IPC, applicants, countries can be done. These new groups being able to be analyzed separately.

These analysis, because they are done almost instantly, provide to the expert all the facilities to develop his own information and to focus on the most relevant information. The full text of the patents can be downloaded if necessary, as well as the content of the local database in different formats which can be integrated in more powerful software or in mailing or cooperative platforms.

If the users have other patent sources, Matheo-Patent can also work on the two US Patent databases and also import data from Delphion (7), Patbase (8) , Micopatent (9) and Derwent (10).

2.1.2 Method

Because the patent source is free, the user can perform as many request that he likes. Then various strategies are available: make a first search and detect significant words, or

IPC, or Applicants or inventors, etc. and complete the database by other queries, or develop separate databases, etc.

The method used is related to the classical cycle of intelligence used many times in the systems of Competitive Intelligence. From the vision, (initial subjects), there will be a selection of information or databases (this is not the case here since we will work on a unique patent database, but the same can be done on the US patent databases if necessary on local home-made databases by using the Matheo Analyzer software). The section of the information being done, the system will provide the facility to manage and handle this information and to perform APA . The results will be either give to an expert o ra group of experts, or the experts may use themselves the system to perform the analysis, correlations, matrix, etc.. the most suitable for them. The objective of the cycle if to provide to the decision makers, here the researchers or the person in charge to coordinate the research the recommendations, alert indexes and available from the implicit information developed during the analysis.

2.1.2.1   *The vision*

The carbon capture and sequestration (CCS) is one of the technologies that is been studied to reduce the CO_2 emission by capturing the CO_2 and storing it in geological formations, using it in enhanced oil recovery (EOR) or in other industrial processes. Coal power plants using the CCS technology is seen by IEA as an important process to reduce CO_2, but it increased the energy production costs by reducing the power plant efficiency.

There are three technologies that can be used to capture the CO_2 in coal power plants:

- Post-combustion: Amine is used to absorb the CO_2 from the flue gas and can easy retrofit existing power plants. It is not an efficient process.
- IGCC (Integrated Gasification Combined Cycle): Coal is gasified producing a syngas containing CO, CO_2 and H_2. CO is then transformed in CO_2 and separated from hydrogen, before it is combusted in a gas turbine. This technology requires a high capital cost during the construction, has a complicated operation and cannot be used to retrofit old power plants.
- Oxyfuel (Oxy-Combustion): The nitrogen is separated from the air and the coal is combusted with a mixture of O_2 and flue gas. In the end of the process, the flue gas contains only water vapor and CO_2, making it easy to separate the CO_2.

From the above consideration there is a need to have some steels which can have a strong resistance to corrosion in a medium close enough of the operation conditions. One of the possibilities is to refer to the following conditions: oxyfuel power plants boilers containing large amount of CO_2 and small amount SO_2 and water at a temperature close from 600ºC.

2.1.2.2 The query

Before the development of various query's strategies, it is important to understand the mechanism of the APA (bibliometrics treatments). The objective is to get from and explicit information (here patent bibliographic data) an implicit information obtained by various statistical treatments. To get this implicit information, the number of data analyzed should be important and around the thematic that we want to explore. Then , the formulation of the query is far away from the classical documentation, where the people try to get the best precise query to obtain only the answer they wish. In this condition, the query pre-determine the answer and to treat the corpus via bibliometrics systems is useless. This is why in the following parts of this presentation we will provide different queries (and their APA results), to familiarize the reader with different approaches not being directly centered on the core subject, but being able to bring to this subject side works and applications from which an implicit information may be obtained.

To illustrate the method, we will present here two query related to the above research.

- Combining IPC (International Patent Classification) (11) and words in title or abstracts: IPC=C22C38/22 and in titles or abstracts steel and (pipe or tube or tubing) from 1940 to 2012 (March)
  We assume here, that one of the most useful corrosion resistant steel is used for pipes and tubes (petroleum, boilers, etc.). this is one way to collect the data from this subject and to test the validity of our assumption.
  IPC C22C38/22 correspond to ferrous alloys (eg steel alloys) and /22 correspond to with molybdenum or tungsten . The use of the IPC is important since it gives the opportunity by varying the choice of the last IPC extension /22 or other to selected various types of alloys. The following extract of the IPC indicates some of these possible selection (we here indicate only one /22 but other queries have also been done and are not presented because of the confidentiality of the subject)

7

→ containing silicon	C22C38/02
→ containing manganese	C22C38/04
→ containing aluminium	C22C38/06
→ containing nickel [N. (C22C38/10B takes precendence)]	C22C38/08
→ containing cobalt	C22C38/10
→ [N. containing Co and Ni]	C22C38/10B
→ containing tungsten, tantalum, molybdenum, vanadium, or niobium	C22C38/12
→ containing titanium or zirconium	C22C38/14
→ containing copper	C22C38/16
→ containing chromium	C22C38/18
→ with copper	C22C38/20
→ with molybdenum or tungsten	C22C38/22
→ with vanadium	C22C38/24
→ with niobium or tantalum	C22C38/26
→ with titanium or zirconium	C22C38/28
→ with cobalt	C22C38/30
→ with boron	C22C38/32
→ with more than 1.5% by weight of silicon	C22C38/34
→ with more than 1.7% by weight of carbon	C22C38/36
→ with more than 1.5% by weight of manganese	C22C38/38
→ with nickel	C22C38/40
→ with copper	C22C38/42
→ with molybdenum or tungsten	C22C38/44
→ with vanadium	C22C38/46

Table 1 Examples of possible selections using the full extended IPC (12)

- Titles = steel and Titles or abstracts Cr and Co and W and Si and Mo and Mn
  General query only linked to the description provided in the patent by the redactor of the demand.

2.2   The results

For the purpose of clarity we will indicates the main results only, because they will underline the general trend of the methodology. All the details will not be presented, since many correlations and analysis can be performed according the experts experiences. This means that the best way to get the most benefit part of this APA is to perform some general correlations (which can be done by non experts), and after to move to refinements done in the presence of one or several experts. Since patents references can be annotated , it is possible to have one expert making a comment and other experts making their own analysis and providing other choices and further comments. In this conditions, experts can work separately with the local database.

2.2.1   General information about the visualization of the Matheo Patent screen

The frequency data which will be used during the presentation is the frequency family. A patent family is the group of patents with a different patent numbers which cover the same invention but for different countries. Only one patent will be indicated when a real family exists, if there is only one patent it will appear as a family with only one member.

In the Matheo Patent system (12), if you click on one item in the left part of the screen (in this case a date) all the patent family afferent to this date will appear on the left upper part of the screen. Selecting one patent in this list will give access the bibliography of this patent. This is general for all the automatic selections appearing on the left (IPC, countries, applicants, inventors, dates, words from titles and abstracts, drawings, or groups of patents selected by the users.

In some cases not all the information is provided by the database, in this case the content related field will not be available.

2.2.2 Different treatments available

The objective is to get most of the information available in the group of patent selected answering the question: who is doing what, where, when, with whom, what are the competency of the applicants, their links, etc. This information can be used to select right patents, to see the trend in technology, eventually the innovative orientations, and to help the users to implement the SWOT analysis or the Porter's diamond.

SWOT : Strength, weaknesses, opportunities, and threats

3.2 IPC=C22C38/22 and in titles or abstracts steel and (pipe or tube or tubing) from 1940 to 2012 (March)

The total patents retrieved is 1045, and the results are presented in family frequency.

2.2.2.1 Chronology.

This is important to see if the number of families or patents is increasing, remains on a plateau or declined. In this domain the number of families is increasing, which indicated that the subject is important but do not give rise to drastic changes from one year to another.

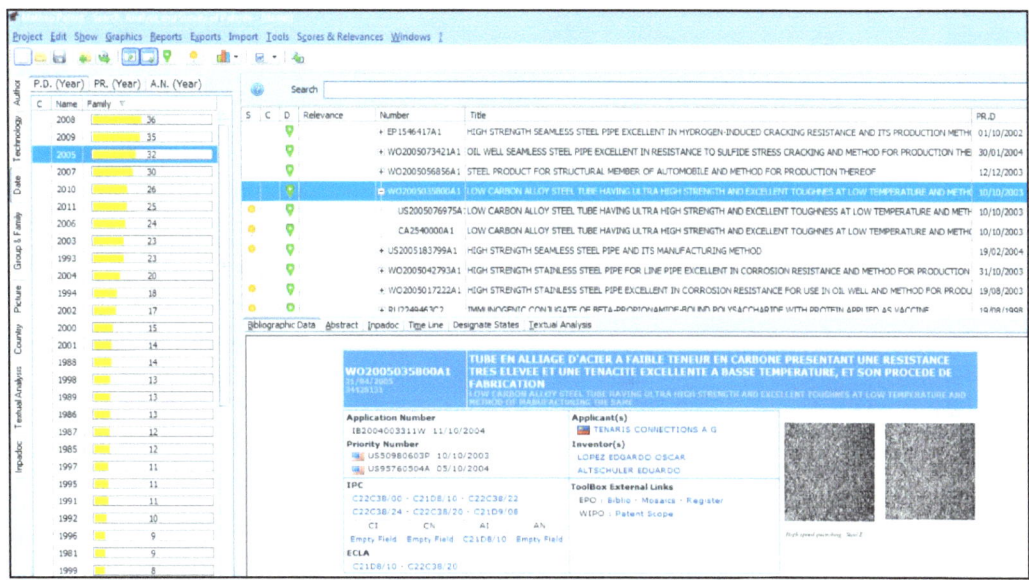

Table 2 **Chronology of the patent granted**

In the bibliographic reference there are also other elements such as:

The equivalent patents:

- AR060286A1 - AT541060T - BRPI0415340A - BRPI0709458A2 - AR046543A1 - WO2007113642A3 - WO2007113642A2
- US2009101242A1 - US2006169368A1 - US2005076975A1 - MX2008012810A - KR20090013769A - KR20060130551A - JP2009532584A
- JP2007508452A - EP2007914A2 - EP1678335B1 - EP1678335A1 - CN101448966A - CN100460527C - CN1902330A
- CA2650452A1 - CA2540000A1

A map is also available in the designated states box following the bibliographic data. This is presented in the following figure.

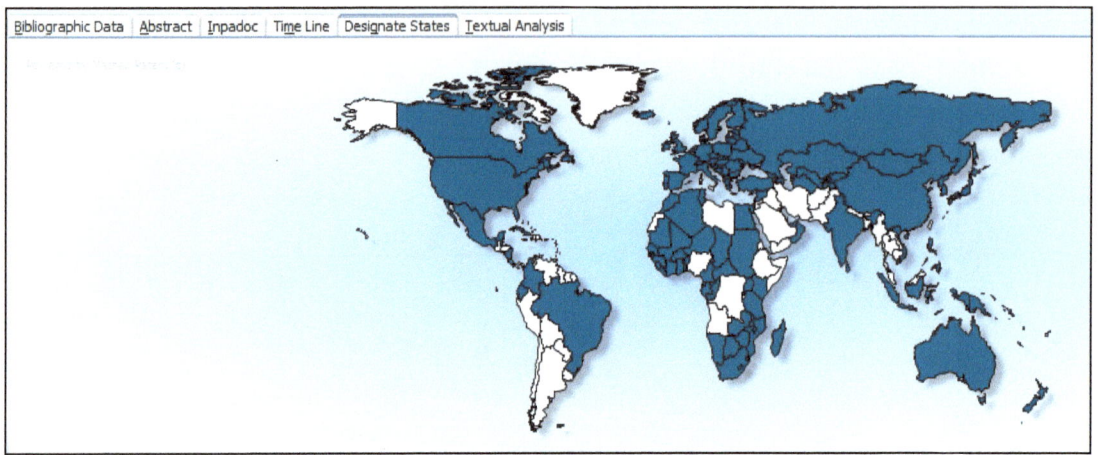

Figure 1Map of the designated states. Note the extension to Africa.

The non patent documents cited by the examiner (these elements are available in US patent, World patents and EP (European) Patents):

Non-Patent literature (13)
- PATENT ABSTRACTS OF JAPAN VOL. 1998
- PATENT ABSTRACTS OF JAPAN VOL. 017
- NO. 10
- 31 AUGUST 1998 (1998-08-31) -& JP 10 140283 A (SUMITOMO METAL IND LTD)
- 26 MAY 1998 (1998-05-26)
- NO. 12
- - NONE
- PATENT ABSTRACTS OF JAPAN VOL. 1999
- PATENT ABSTRACTS OF JAPAN VOL. 016
- NO. 244 (C-0947)
- 4 JUNE 1992 (1992-06-04) -& JP 04 052225 A (SUMITOMO METAL IND LTD)
- 20 FEBRUARY 1992 (1992-02-20)
- 29 OCTOBER 1999 (1999-10-29) -& JP 11 199929 A (SUMITOMO METAL IND LTD)
- 27 JULY 1999 (1999-07-27)
- NO. 055 (C-1023)
- 3 FEBRUARY 1993 (1993-02-03) -& JP 04 268016 A (KOBE STEEL LTD)
- 24 SEPTEMBER 1992 (1992-09-24)

Cited patents (patents related to the invention cited by the examiner)(13)
- JP10140283A - US2003155052A1 - JP11199929A - US5454883A - JP4052225A - JP4268016A

2.2.2.2　The main technologies

The analysis of the IPC (note that a patent may be indexed with more than one IPC) allows to determine what are the most important technologies involved. This is done automatically. There are two ways to analyze the technologies:

- A list of the main IPC with their frequencies

The table indicates the most important iPC 4 digits

IPC codes	IPC signification	Frequencies
C22C	ALLOYS	185
C21D	MODIFYING THE PHYSICAL STRUCTURE OF FERROUS METALS; GENERAL DEVICES FOR HEAT TREATMENT OF FERROUS OR NON-FERROUS METALS OR ALLOYS; MAKING METAL MALLEABLE BY DECARBURISATION, TEMPERING, OR OTHER TREATMENTS	103
B21B	ROLLING OF METAL	32
E21B	EARTH OR ROCK DRILLING; OBTAINING OIL, GAS, WATER, SOLUBLE OR MELTABLE MATERIALS OR A SLURRY OF MINERALS FROM WELLS	20
F16L	PIPES; JOINTS OR FITTINGS FOR PIPES; SUPPORTS FOR PIPES, CABLES OR PROTECTIVE TUBING; MEANS FOR THERMAL INSULATION IN GENERAL	18
B21C	MANUFACTURE OF METAL SHEETS, WIRE, RODS, TUBES OR PROFILES, OTHERWISE THAN BY ROLLING; AUXILIARY OPERATIONS USED IN CONNECTION WITH METAL-WORKING WITHOUT ESSENTIALLY REMOVING MATERIAL	17
B23K	SOLDERING OR UNSOLDERING; WELDING; CLADDING OR PLATING BY SOLDERING OR WELDING; CUTTING BY APPLYING HEAT LOCALLY, e.g. FLAME CUTTING; WORKING BY LASER BEAM	13

Table 3 **Most important IPC, their frequencies and signification**

Using the list of IPC, it is possible to build different groups of patents. We choose to build the group afferent of the IPC C22C dealing with alloys. Doing this, we separate the patents dealing with alloys to other patents not necessary relevant with the search (for instance patents dealing with a certain boiler's shapes, etc.).

Starting from this group, it is possible to develop different analysis such as main applicants, benchmarking of the companies (matrix IPC4, Applicants), networks of Applicants (links between companies), etc. In fact all the fields presents in a patent reference can be used to build up lists, matrix, networks . The figure indicates part of the data which can be used to build various matrix

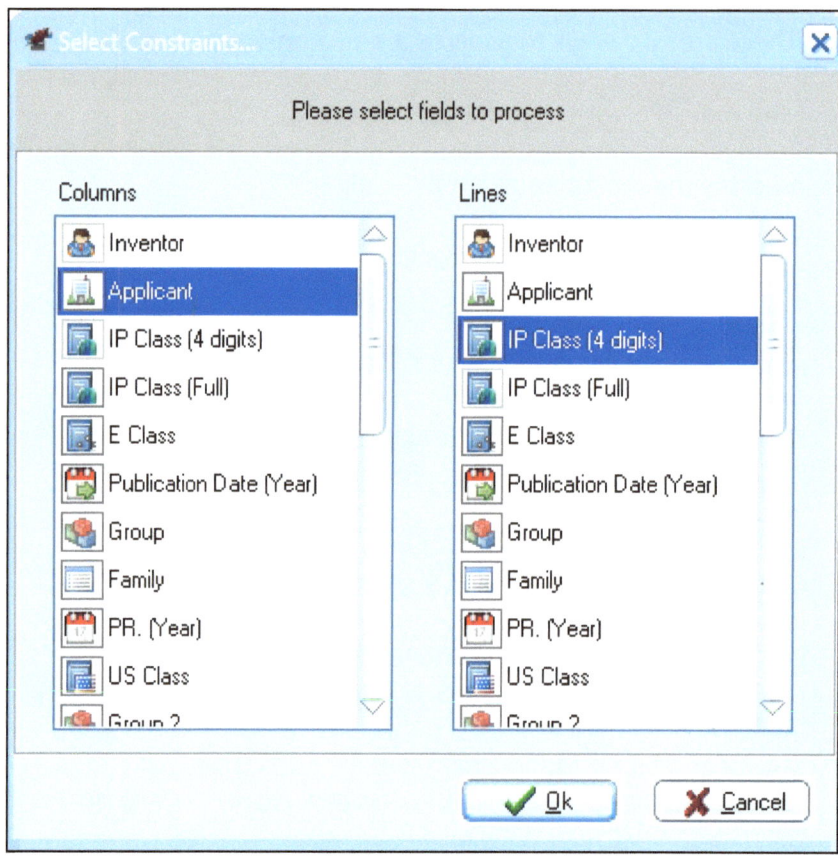

Figure 2 How to build a matrix

2.2.2.3   Main applicants present in the C22C IPC

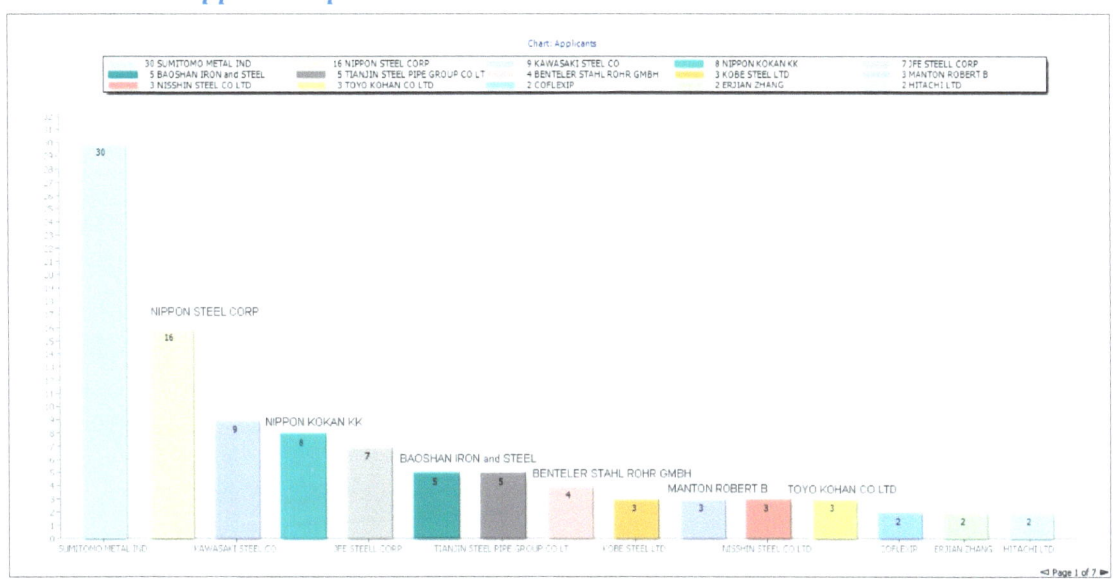

Figure 3 Main Applicants in the C22C group of patents

2.2.2.4   Benchmarking of the competencies of the main applicants (above). This is done by making a matrix between the applicants and the IPC

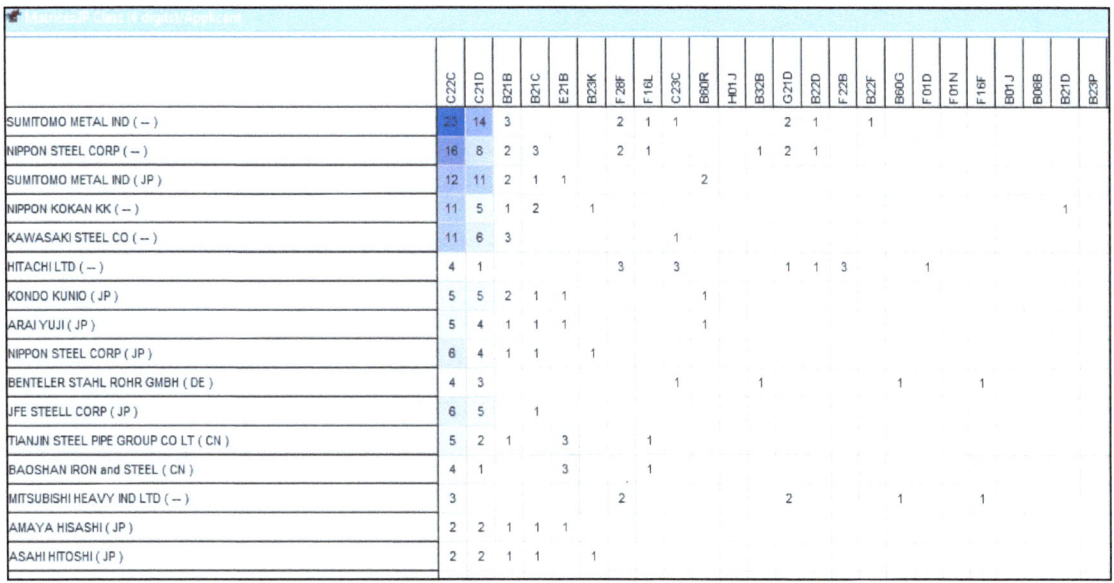

	C22C	C21D	B21B	B21C	E21B	B23K	F28F	F16L	C23C	B60R	H01J	B32B	G21D	B22D	F22B	B22F	B60G	F01N	F01N	F16F	B01J	B08B	B21D	B23P
SUMITOMO METAL IND (--)	23	14	3			2	1	1						2	1	1								
NIPPON STEEL CORP (--)	16	8	2	3		2	1					1	2	1										
SUMITOMO METAL IND (JP)	12	11	2	1	1					2														
NIPPON KOKAN KK (--)	11	5	1	2		1																	1	
KAWASAKI STEEL CO (--)	11	6	3						1															
HITACHI LTD (--)	4	1				3		3				1	1	3		1								
KONDO KUNIO (JP)	5	5	2	1	1				1															
ARAI YUJI (JP)	5	4	1	1	1				1															
NIPPON STEEL CORP (JP)	6	4	1	1	1																			
BENTELER STAHL ROHR GMBH (DE)	4	3							1			1				1		1						
JFE STEELL CORP (JP)	6	5	1																					
TIANJIN STEEL PIPE GROUP CO LT (CN)	5	2	1		3			1																
BAOSHAN IRON and STEEL (CN)	4	1			3			1																
MITSUBISHI HEAVY IND LTD (--)	3					2								2				1		1				
AMAYA HISASHI (JP)	2	2	1	1	1																			
ASAHI HITOSHI (JP)	2	2	1	1		1																		

Figure 4 Benchmarking of the main applicants (inpart)

2.2.2.5 Analyzing in detail the local database using the selection of significant words extracted from the tittles and abstracts.

The selection of patents related to some words or here chemical elements for instance, can also be combined with a direct search on the selected patent. An example of this combination is presented in the following figure.

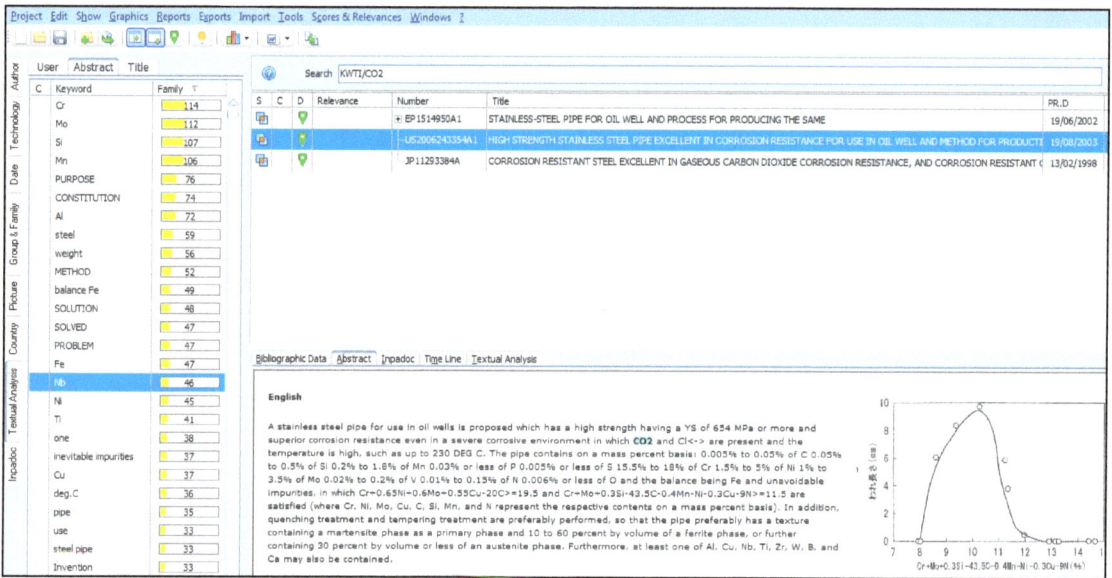

Figure 5 election of the families

Selection of the families containing the chemical Nb (Niobium) and on this selection, serch for patents containing CO2 (Carbon dioxide). The abstract of the patent selected from this title (bleu selection), is presented at the bottom part of the screen.

Another example is show in figure 8. From the list we selected the following terms: ferritic, martensic, austenitic, ferrite-ferritic, rare earth and build up a groups of patents for each of them. Various analysis may be performed on these group, lest, matrix, etc…The figure 7 represents the matrix made with the above groups and other groups of priority countries (first patent of the invention granted in this country).

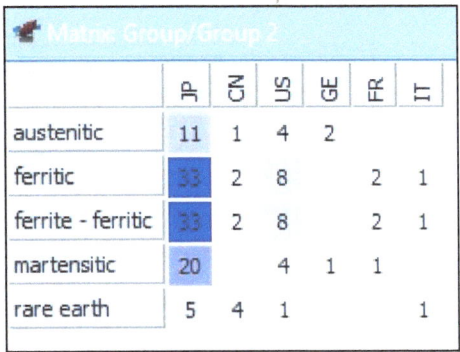

	JP	CN	US	GE	FR	IT
austenitic	11	1	4	2		
ferritic	33	2	8		2	1
ferrite - ferritic	33	2	8		2	1
martensitic	20		4	1	1	
rare earth	5	4	1			1

Figure 6 Specialization by country

Another way to explore the information downloaded in the local database, is to perform offline queries on this corpus. This will allow the retrieval of patents dealing with specific data: for instance the search for steels containing chromium (Cr) and cobalt (Co). The result is presented in figure 9.

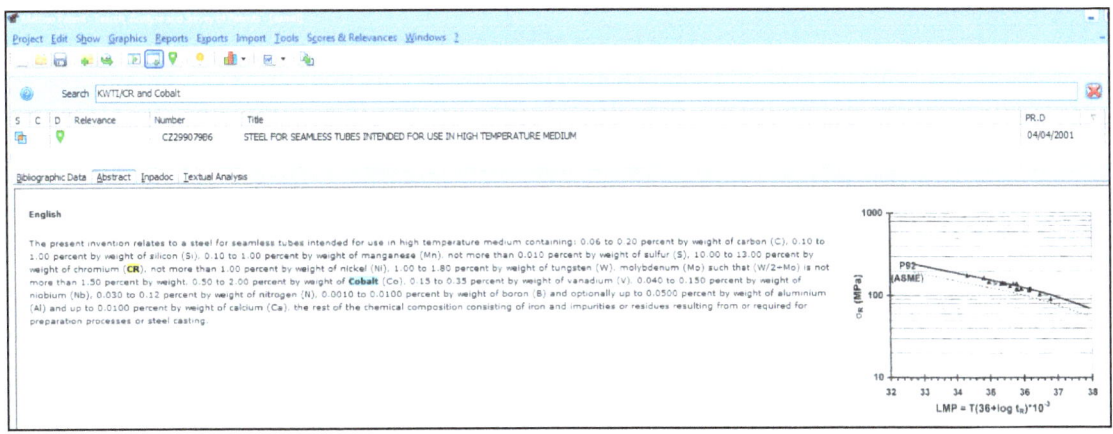

Table 4 Selection of patent with Chromium and Cobalt

2.2.3 Titles = steel and Titles or abstracts Cr and Co and W and Si and Mo and Mn

The process is the same than for the former analysis. After the query, the selected patents are uploaded and a local database is built. Then various analyses may be performed using all the available data from the patent references. 525 patents are selected. We realize various groups (countries, selected words from titles and abstracts, etc.) which will be used in various correlations. We will present some of them as examples.

2.2.3.1 Families per countries (priority countries):

countries	Family frequencies
Japan	222
China	39
Sweden	22
USA	16
South Korea	15
France	9
Germany	8

Table 5 Number of families per priority countries

2.2.3.2 Selection from titles and abstracts words

The figure present various results dealing with the realization of groups from martensitic, ferritic, austenitic and rare earth, and the combination of these groups with other data and groups.

Figure 7 Various examples of analysis

2.2.3.3 The case of the Chinese patents

If we examine the Chinese priority patents we can see that all of them are not extended to other countries (exempts if for the patents with less than one year of age, an extension may be pending). In this condition this means that all the information present in these Chinese patents (there is no family only patent since not extended) can be used without paying any royalty (14). One may also note that some Chinese data are utility models (protection during 10 years, extension to patent within the framework of double patenting is possible[7] if the scope of the protection is different and if the span of protection is also different).

For instance if we search in the database a Chinese patent in the martensic group, we will obtain:

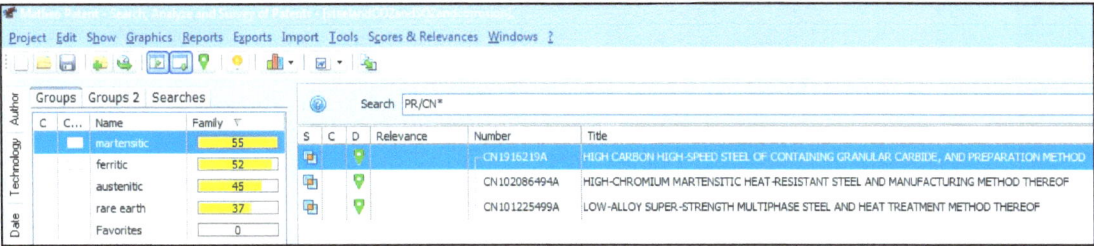

If one of these patents seems particularly important, it is easy to download the full text of the patent from one of the options of the software (document inquiry of Matheo Patent). Here is an example of the first page of the second Chinese patent above:

---

[7] For more information see Stephen Yang is a patent attorney and partner with Peksung Intellectual Property
http://www.google.com/url?sa=t&rct=j&q=&esrc=s&frm=1&source=web&cd=1&sqi=2&ved=0C B4QFjAA&url=http%3A%2F%2Fwww.ipworld.com%2Fipwo%2Ffileserve.htm%3Fdircode%3D1%2 6file%3DMay2012%252F08%252FImpact%2Bof%2Bnew%2Bprovisions.pdf&ei=4kRFUPGkI6TM6 wG3lIHIDw&usg=AFQjCNFY3BeV3RIUGF5OUtdMT2LG1Gp1VA&sig2=8RtVr9GRXpCbC9VGFzuLnw

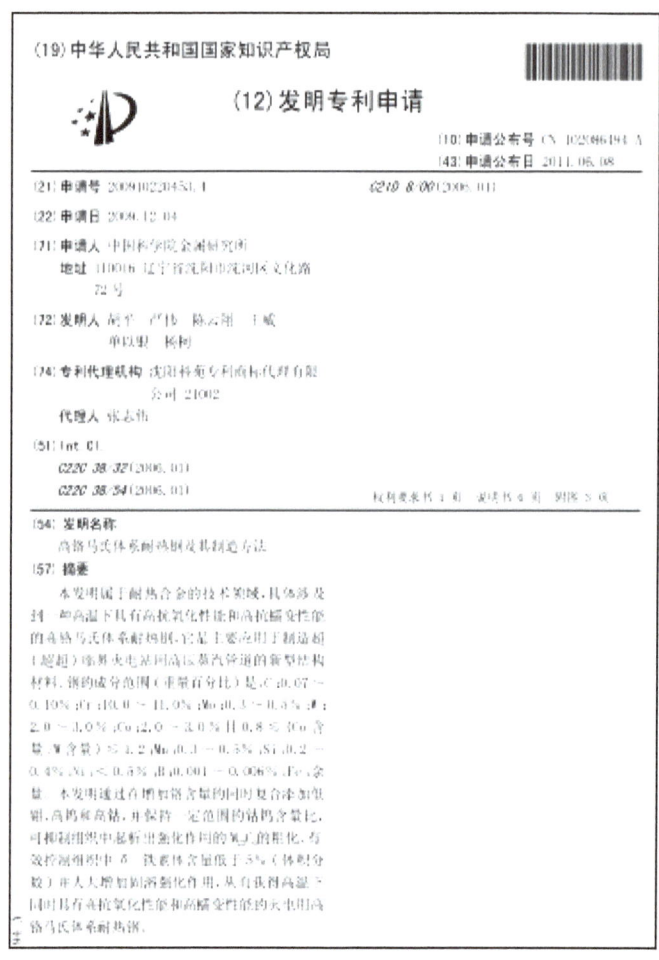

Figure 9 First page of the Chinese patent

The Chinese patents when they are not extended are in Chinese. The problem of translation, can be partially overcome by treating the pdf document with and OCR (for instance using Adobe Professional, and after using a translator such as Sistran (15) or Google translator.

If the bibliographic reference must be included in an email or inserted in a cooperative platform, it is possible to extract one or several references or all the databases in different formats such as text, HTML, etc.

Example:

TI_EN - HIGH-CHROMIUM MARTENSITIC HEAT-RESISTANT STEEL AND MANUFACTURING METHOD THEREOF

PN - CN102086494A

UR   -
http://worldwide.espacenet.com/publicationDetails/biblio?CC=CN&NR=102086494&KC=A&FT=E
&DB=EPODOC&locale=en_EP

PD - 08/06/2011

AB_EN -

The invention belongs to the technical field of heat-resistant alloys, and particularly relates to high-chromium martensitic heat-resistant steel with high oxidation resistance and high creep resistance at a high temperature. The high-chromium martensitic heat-resistant steel is a novel structural material mainly applied to the manufacturing of high-pressure steam pipelines for super (ultra super)-critical thermal power stations. The steel comprises the following components in percentage by weight: 0.07 to 0.10 percent of C, 10.0 to 11.0 percent of Cr, 0.3 to 0.5 percent of Mo, 2.0 to 3.0 percent of W, 2.0 to 3.0 percent of Co (the content ratio of Co to W is more than or equal to 0.8 and is less than or equal to 1.2), 0.3 to 0.5 percent of Mn, 0.2 to 0.4 percent of Si, less than 0.5 percent of Ni, 0.001 to 0.006 percent of B, and the balance of Fe. In the high-chromium martensitic heat-resistant steel, by compoundly adding low molybdenum, high tungsten and high cobalt when the chromium content is increased and keeping the content ratio of cobalt to tungsten within a certain range, the coarsening of M23C6 which plays a role in precipitation strengthening in a microstructure can be inhibited, the content of delta-ferrite in the microstructure is effectively controlled to be lower than 5 percent (volume fraction), and the solid solution strengthening effect is greatly improved, so that the high-chromium martensitic heat-resistant steel which has both high oxidation resistance and high creep resistance at a high temperature and is used for thermal power stations is obtained.

RP - CN102086494A

PR - CN20091220453 20091204

OPR - CN20091220453 20091204

DPR - 20091204

IC - C22C38/32; C21D8/00; C22C38/54

I4 - C22C; C21D

IAI - C22C38/32; C21D8/00; C22C38/54

PA - INST METAL RES CHINESE ACAD SC (--)

IN - PING HU (--); WEI YAN (--); YUNXIANG CHEN (--); WEI WANG (--); YIYIN SHAN (--); KE YANG (--
)

The analysis, may be continued with experts by developing new correlations, lists, matrix, networks, etc. to give the best possible knowledge of the subject. If necessary other queries can be done and the local database can be updates or new databases can be created.

2.3 Conclusion

The use of APA (Automatic Patent Analysis) to know the technical situation of a subject is particularly interesting because patent documents make a bridge between fundamental research and applications. The information available in a patent reference allows after a bibliometrics treatment to answer various questions about the main applicants (companies), the new possible entrants, the chronology of the inventions related to the subject, the search for partnerships, the identification of the main competitors, etc. This is a good assistance to provide information for the Porter's Diamond or the SWOT analysis. Today, the development of innovation, that is to say the use of competences to create new products or services leads directly to PPP (Public and Private Partnerships) (16)and the Triple Helix (17)   the APA is a good way to develop or induce such partnerships.

It is also very important to keep in mind in these analysis that what we see (during the interval of time choose for the search) is what the people did in the past or in the recent past. But not what do they do today. Then, if the analysis indicates companies and inventors dealing with research or applications directly linked to the subject, they may be considered as "targets" to be followed to see in which direction they work today using the classical scientific databases (from Dialog[8] for instance), the Internet and web sites, the international meetings and workshops, etc.

In these examples, we work with the world patent database, available from EPO (European Patent Office), this database covers more than 80 countries. This database index only the titles and the abstracts of the patents, but not the full text of it. If searches want to be done on the full text of the patents this can be done for the US patents by using Google Patent. Google index all the text. This can be valuable to make "string searching", but if you use only separate words you may get many artifacts.

2.4 Bibliography

1 – http://www.matheo-softaware.com –consulted April 2012). A trial version if available online and can be downloaded.

---

[8] http://library.dialog.com/bluesheets/

21

2 – Worldwide, full collection of patent published applications from more than 80 countries, http://worldwide.espacenet.com/advancedSearch?locale=en_EP (consulted April 2012)

3 – EPO European Patent Office, http://www.epo.org/ (consulted April 2012)

4-- Henri Dou, Valérie Leveillé, Sri Damayanty Manullang, Jean-Marie Dou Jr, Patent Analysis for Competitive Technical Intelligence and Innovative Thinking, Data Science Journal (DSJ), Vol. 4 (2005) pp.209-236 https://www.jstage.jst.go.jp/article/dsj/4/0/4_0_209/_article

5 – Tugrul U. Daim, Guillermo Rueda, Hilary Martin, Pisek Gerdsri, Forecasting emerging technologies: Use of bibliometrics and patent analysis, Technological Forecasting and Social Change, Volume 73, Issue 8, October 2006, Pages 981–1012

6 –Hervé Rostaing, La bibliométrie et ses techniques, Sciences de la Société, Collection "Outils et Méthodes", 131 p., 1996

7 - http://www.delphion.com/ (consulted April 2012)

8 - http://www.wipo.int/export/sites/www/aspi/fr/doc/patbase_summary.pdf (consulted April 2012)

9 - http://www.micropat.com/static/index.htm (consulted April 2012)

10 - http://library.dialog.com/bluesheets/html/bl0351.html (consulted April 2012)

11 - http://worldwide.espacenet.com/eclasrch?classification=ecla&locale=en_EP&ECLA=C09D5 (consulted April 2012)

12 – Henri Dou, Benchmarking R&D and companies through patent analysis using free databases and special software: a tool to improve innovative thinking, World Patent Information, Volume 26, Issue 4 , December 2004, Pages 297-309

13 – M.M.S. Karki, Patent citation analysis: A policy analysis tool, World Patent Information, Volume 19, Issue 4, December 1997, Pages 269–272

14 – WIPO World International Property Organization, http://www.wipo.int/patentscope/en/ (consulted April 2012)

15 – Automatic translation tools, http://www.systran.fr/ (consulted April 2012)

16 – Jacky Kister and Henri Dou, Integration of Competitive Intelligence and Watch in Academic Scientific Research Laboratory (Chapter), Competitive Intelligence and Decision Problems, Edited by Amos David, ISTE, WILEY, 2011, ISBN 978-1-84821-237-4, pp225-242

17 - Leydesdorff L., Etzkowitz H., " The Triple Helix as a Model for Innovation Studies, (1998), (Conference Report)", Science & Public Policy Vol. 25(3), pp.195-203
See also i Industry & Higher Education (1998), 12, n° 4, pp. 197-258
http://en.wikipedia.org/wiki/The_Triple_Helix   (consulted April 2012)

18 - Rajendra Prasad, ReviewofF Soil Remediation Technbologies trhough Patent Analysis, International Conference on Environmental Exposures in the Era of Climate Change, 20-22, November, 2009 : Perth, Australia - 13th Int Conf of The Pacific Basin Consortium For Environment & Health,
http://s244543015.onlinehome.fr/ciworldwide/?p=698 (consulted April2012)

3   Analyse de l'Information Brevet comme méthode pour accroître le dialogue entre la recherche académique et l'industrie. Une voie pour favoriser la pre-clustérisation et le développement de projets.

Henri Dou (*)

(*) Directeur ATELIS (Atelier d'Intelligence Stratégique - France Business School - fBS)), France. douhenri@yahoo.fr www.atelis.org   www.ciworldwide.org

Résumé: les brevets sont considérés par la majorité des gens comme un système de protection du capital intellectuel. Mais, ceci est seulement une facette de leur utilisation. L'APA (Analyse Automatique des Brevets) permet d'utiliser l'information technique provenant de plus de 90 ans et représentant plus de 60 million de références. Ceci permet de positionner un sujet dans le cadre du développement technologique et mettre en évidence les éléments d'information agrégés utiles pour le développement stratégique : nouvelles applications, partenaires potentiels, etc. Dans cette présentation nous présentons un cas réel focalisé sur l'aquaculture. Cet exemple va montrer comment une information adaptée devient attractive pour les PME et ETI mais aussi pour les chercheurs académiques. Cela permet de développer de nouveaux liens entre les parties prenantes d'un même domaine, et ainsi de favoriser le développement. Nous utiliserons la base de donnée mondiale de l'Office Européen des brevets (OPS VI) pour extraie ls données afférentes à l'aquaculture, en utilisant comme système de travail le logiciel résident Matheo-Patent. Ceci va permettre de développer une base de donnée pre-codée qui va permettre de réaliser, par combinaison des divers champs documentaires présents dans une notice de brevet, toutes les analyses possibles (listes, matrices, réseau, ...). L'étude qui est présentée ici concerne la période de 2000 à Septembre 2012. La sélection des brevets a été réalisée en utilisant simplement le terme « aquaculture » dans le titre et le résumé. Il est évident que pour une étude exhaustive d'autres termes pourraient être utilisés de même que des combinaisons entre certains termes et la classification internationale des brevets. On accède ainsi à une information élaborée (« benchmarking » des compétences des inventeurs, des déposants, évolution du domaine dans le temps, possibilité de détecter de nouveaux entrants ainsi que des orientations innovantes. Ceci va permettre la réalisation de tableaux de bord facilitant la pre-clustérisation des parties prenantes ainsi que la mise en place d partenariats publics privés (PPP).

Mots clés: : Intelligence Compétitive, Intelligence Economique, Innovation, Brevets APA (Automatic Patent Analysis), Développement régional, Bibliométrie, partenariat public privé, cluster, aquaculture

Abstract: patents are for most users a tool for protecting intellectual property. But this is only a facet of their use. The APA (Automatic Patent Analysis) allows the use of patent information (over 90 million records covering more than 90 countries) to provide strategic elements (matrices, networks, lists, etc.) which allow to position a subject in the context of technologies and applications, highlighting new applications, potential partners, etc. In this paper we present a real case study which focuses on aquaculture This example will show how the right information may attract the SMEs interest as well as the one of academics, developing new links between all the stakeholders of this area and promoting project development.. We used the EPO worldpatent database (OPS-VI) to extract the data (aquaculture and also sea fish breedind) using as a tool the Matheo-Patent software (to develop the local database and to perform the APA (Automatic Patent Analysis)). The time period used was from 2000 to present (September 2012). In this case study, the various applications are described, the strategic relationship between applications, applicants (industries) are examined as well as the necessary technologies for the realization of certain products. These results will provide a dashboard for SMEs and decision makers to develop a pre-clusterization as well as to promote PPP (Private and Public Partnerships). In this presentation we use only the word aquaculture as research element in the titles and abstracts of the patents. It is obvious that for an exhaustive search other words may be used (boolean OR) and that combinations or use of the CIB is also possible.

Key words: Competitive Intelligence, Patent, APA (Automatic Patent Analysis), Regional Development, Bibliometrics, PPP Public and Private partnerships, Cluster,, Aquaculture,

3.1 Introduction

3.1.1 Innovation

The development of innovation most of the time is done in two steps: (Erikson 2006)

The first one is the development of competences and knowledge in Universities and Research centers mainly with the financial support of the Government.

The second one is at least also important than the first, is to transform these knowledge and competences in products and services with the help of the industry.

This classical model[9] gives rise to clusters and in France to the poles of competitiveness where the intersection of the Research, States Institutions and Industry provides the best conditions to achieve the second step. Taking the problem that way means that to develop a cluster it is necessary to go from a scattered state of actors to a pre-clusterized one. Different works have been done about this subject, but we believe by experience, that to attract the attention and concern of the stakeholders it is necessary to create a strong incentive and to show that the sector (which can be economic, industrial, touristic ...) will benefit from this experience. It is then necessary to find "something" which can bridge the gap between research, state institutions and industries.

3.1.2 Patents

Patents, which open a large window on applications and products development (Yanhong 2007), (Guellec 2001), (Zoltan 1998) can be used in this case, not to protect an invention, but to provide a way to underline what others (university , industry, research centers, individuals...) develop here in the aquaculture and sea fish breeding fields. It is also interesting to see that what is published in patent is most of the time not published elsewhere and also that patents are practically never cited in academics bibliography. Then the patent analysis dealing with a specific field prefigures a virtual cluster which, for the local stakeholders will prefigure also what can be locally developed. Various aspects of the use of text mining or APA or biblbiometrics (Rostaing 1996) provide examples of their potential aggregating power.

---

[9] We exclude in Innovation the American Model (high level of stratup and risk capital, as well as Japan's one where most of the research is done in powerful industrial firm with low relationship with university.

3.1.3 Competitive Intelligence

Patents can be used in Competitive Intelligence to protect the Intellectual Property assets but, they are also a living technical encyclopedia which may provide insight on the development of various areas of science and technology. It is well know that it is difficult for SMEs to understand the language and uses of the results presented in academics journals. In the same time it is difficult for the scientist to get an overview of what can be done in the industrial sector from the basic knowledge and the competences developed in his laboratory. To fill this gap, the patent information is a must since it provides words, drawings claims and operatory modes, which can be understood by both parties. Moreover, if you look to the references provided in academics papers, you will note that patents are almost never cited. This is why the use of APA, is a powerful method to provide to both parties understandable information which will help to build up a dialog between all the stakeholders of a specific area of science and applications. The combination of academics and patents information has been one of the best ways to promote new links between science and technology and in certain case to induce a pre-clusterization and project developments.

3.2 Material and method

3.2.1 Materials

We use as an information source the database of EPO (European Patent Office OPS-VI) (Espacenet 2012) which allows the query of the database by robots. The robot, here is the Matheo-Patent software (Matheo 2012) allows to query the database, to build up a local formatted database and to perform automatically the APA (Automatic Patent Analysis). Various papers have already been published and they described the main functions of the above software (Dou 2005, 2004)

3.2.2 Methodology

The query of the database is done by using the **words aquaculture for one search and fish and breeding for the other.** If necessary the query can be performed on titles and abstracts or only in titles. Because this presentation emphasizes the methodology, it is also possible in some cases to use the International Patent Classification (Espacenet 2012) to get the best possible selection. When the search is finished patents or a selection from the patents can be downloaded to build up a local formatted database. This database can be updated in time or the result of new searches can be added if necessary. The objective is to get the most relevant patent selection according the subject and the expert point of view.

All analysis and correlations (lists, matrix, networks) which can be built from the documentary fields available in the downloaded patents (titles and abstracts words, Patent dates, number, priority, IPC (International Patent Classification from 4 or 8 digits) (IPC 2012)

Creation of patent groups according a technology (IPC or significant expression of titles and abstracts), dates, countries, etc.

Moreover on each group or on various parts of the local database the following treatments will be available:

- All analyses and correlations (with the documentary fields present)
- Creation for each important patent of performance index and a comment if necessary (this will facilitate the work of different experts on the same set of data)
- Export of any part of the data in text, XML or CSV formats
- Realization of various automatic reports in Word format from the local database (full, IPC, Patent Assignees, etc.)
- The update of the database at any time

Among the patents retrieved a particular attention will concern the patents which are not extended to France or Canada, this is the case for various Chinese patents (Dou 2012). This is because beyond the 12 months of possible extension, all the ideas, detail, etc. described in these patents will be available free (WIPO 2012).

This study can also be extended using the same software to the two databases of the USPTO (US Patent and Trademark Office), patent granted and patent on demand.

3.3   Functions of the Matheo-Patent software

3.3.1   General functions

The main screen after downloading presents the titles of the patents, one mouse click on the title gives access to the full title, abstracts, significant expressions from titles and abstracts, geo-mapping if available, etc. The full text of the patent can also be available using a right click of the mouse on the title. If the title if preceded by the + sign, clicking on the sign will open the patent family members. All these facilities are presented in figures from 1 to 6.

The figures after the figure 6 present most of the automatic analysis which can be performed on the local database or on various groups selected patents.

The local database can also be queried locally on all the available documentary fields (Titles, Abstracts, inventors, Patent Assignees, IPC, etc.)

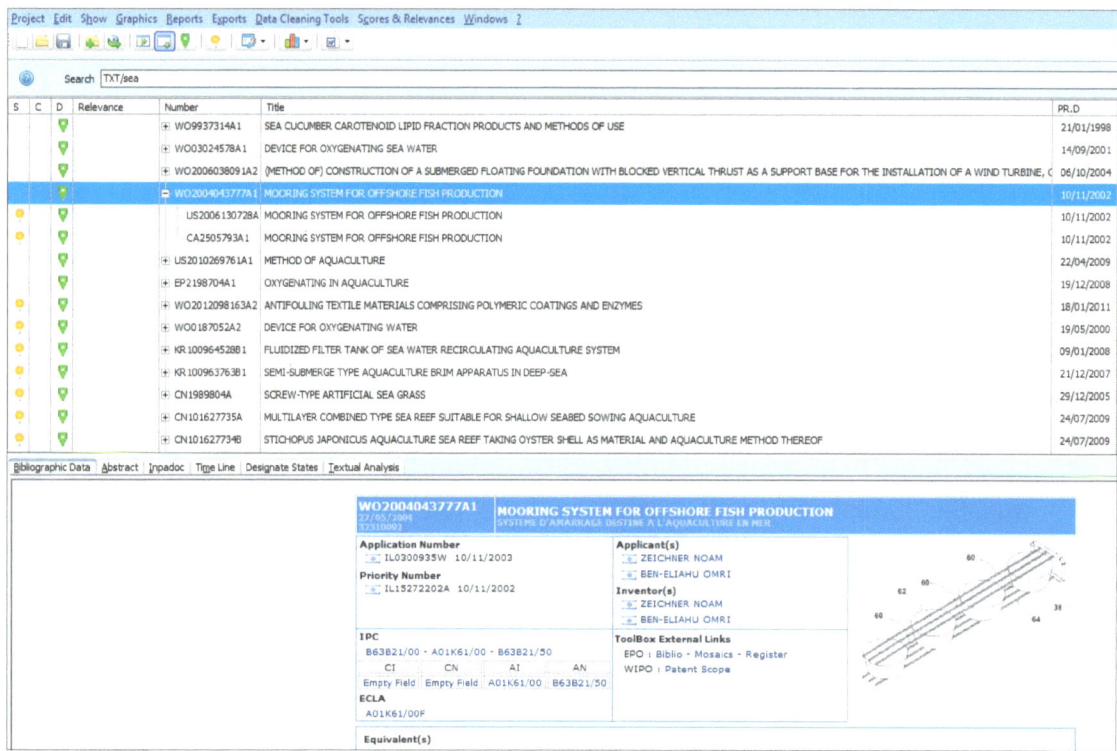

Note that the query of the database was made with the word aquaculture to build up the most relevant local database. In the above screen we selected from this database only the patent which the word "sea" in titles or abstracts. The selected patent (in blue), allows to see the patent family by clicking on the plus sign close from the patent number. The bibliographic data are provided in the lower part of the screen.

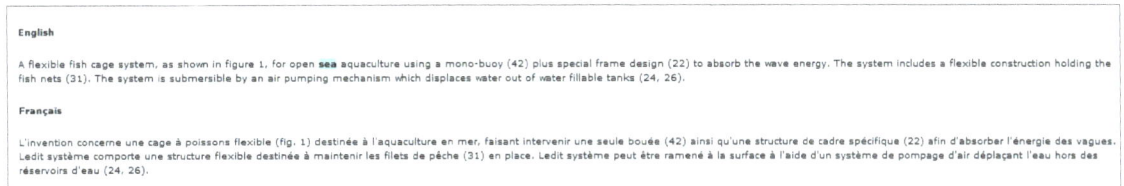

English

A flexible fish cage system, as shown in figure 1, for open **sea** aquaculture using a mono-buoy (42) plus special frame design (22) to absorb the wave energy. The system includes a flexible construction holding the fish nets (31). The system is submersible by an air pumping mechanism which displaces water out of water fillable tanks (24, 26).

Français

L'invention concerne une cage à poissons flexible (fig. 1) destinée à l'aquaculture en mer, faisant intervenir une seule bouée (42) ainsi qu'une structure de cadre spécifique (22) afin d'absorber l'énergie des vagues. Ledit système comporte une structure flexible destinée à maintenir les filets de pêche (31) en place. Ledit système peut être ramené à la surface à l'aide d'un système de pompage d'air déplaçant l'eau hors des réservoirs d'eau (24, 26).

Figure 11 Main screen, access to the abstract (in several languages if available)

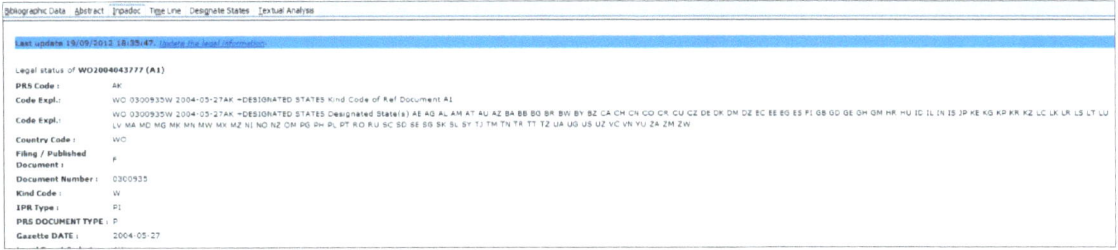

Figure 12 Main screen, Inpadoc data (here in part) if available (Inpadoc 2012)

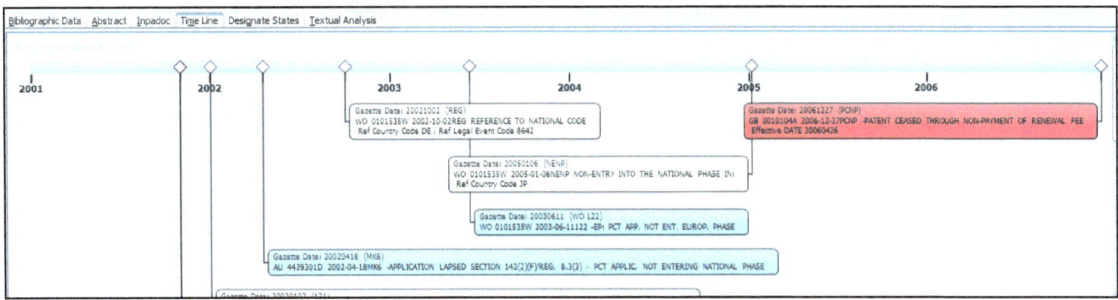

Figure 13 Main screen, access to the « life of the patent » (chronology of events given by the time line)

Here, the information is not relevant to the above patent, since nothing was indicated on the time line, but from another patent to give to the reader a view of the provided information.

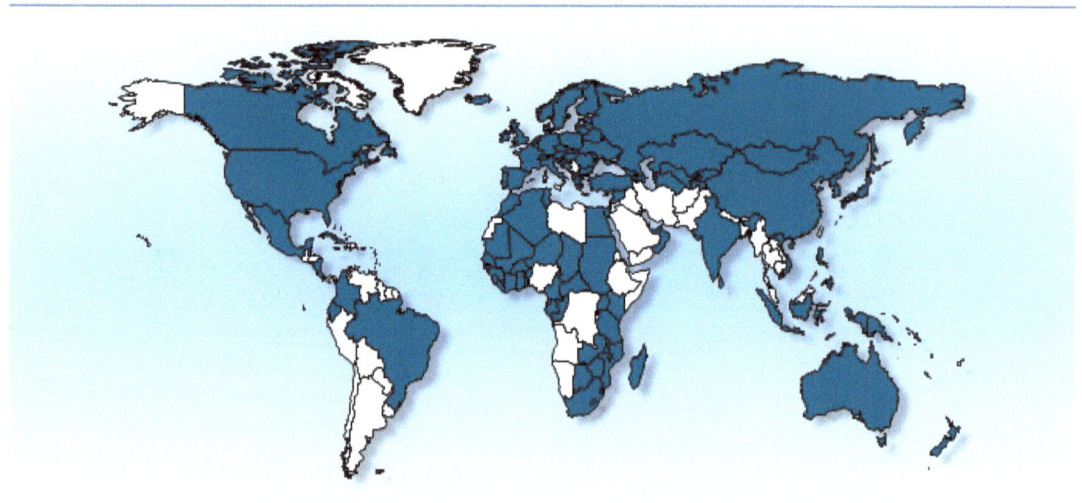

Figure 14 Main screen, mapping of the designated states, patent IL0300935W

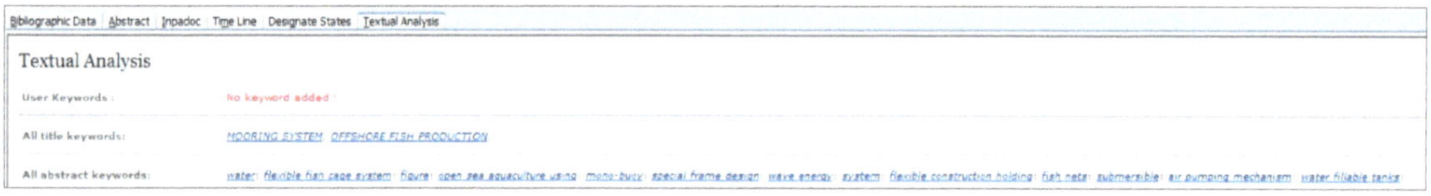

| Bibliographic Data | Abstract | Inpadoc | Time Line | Designate States | Textual Analysis |

Textual Analysis

User Keywords :	No keyword added :
All title keywords:	MOORING SYSTEM · OFFSHORE FISH PRODUCTION
All abstract keywords:	water· flexible fish cage system· figure· open sea aquaculture using· mono-buoy· special frame design· wave energy· system· flexible construction holding· fish nets· submersible· air pumping mechanism· water fillable tanks

Figure 15 Main screen, significant expressions extracted from the title and the abstract (Patent IL0300935W)

Note the presence of the field User keywords which means that a special indexation can be made by the user.

3.3.2 Examples of various analysis

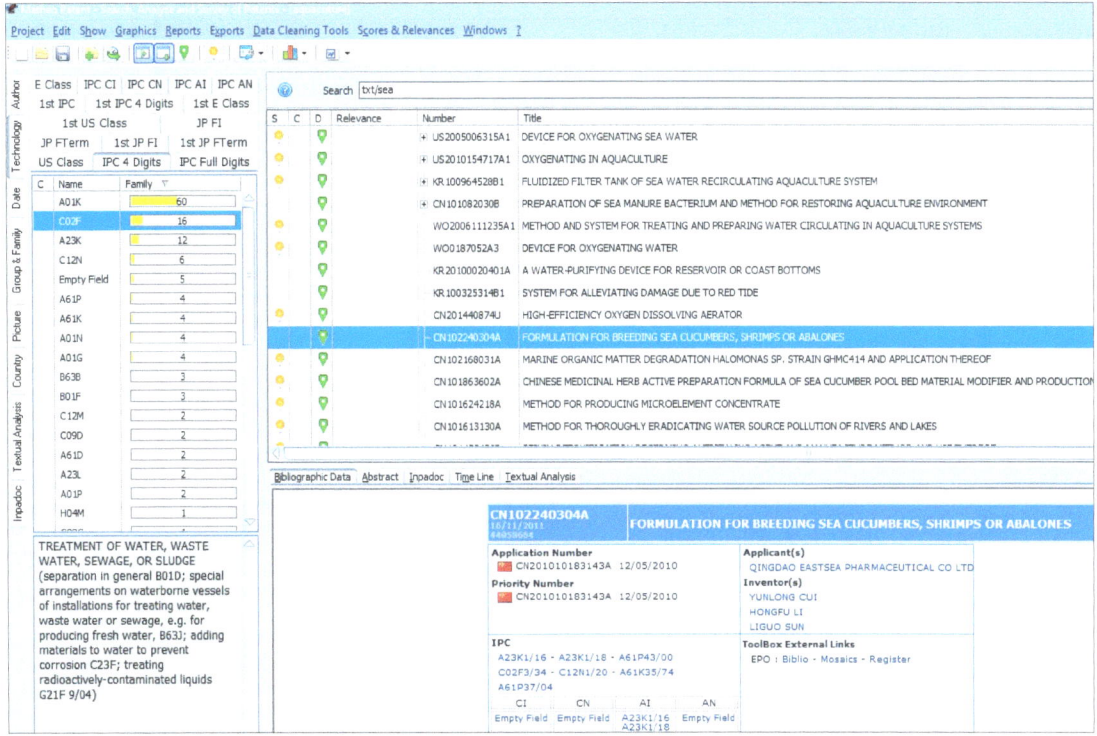

Figure 16 Main screen, significant expressions extracted from the title and the abstract (Patent IL0300935W)

The build in function Show → Patent Analysis, provide instantly the listing of the Applicants, Inventors, IPCs, Patent dates, significant expressions provides from titles and abstracts, drawings and groups (if selected and built up by the user). Here the IPC 4 digits list is provided the IPC C02F (TREATMENT OF WATER, WASTE WATER, SEWAGE, OR SLUDGE) has been selected. Al the patents concerned by this IPC appeared on the left (here follows by a selection of the word sea in titles or abstracts). One Chinese patent has been selected. Note that this patent from 2010 has not been extended out of China. Then, all the date provided in this patent can be used freely in other countries.

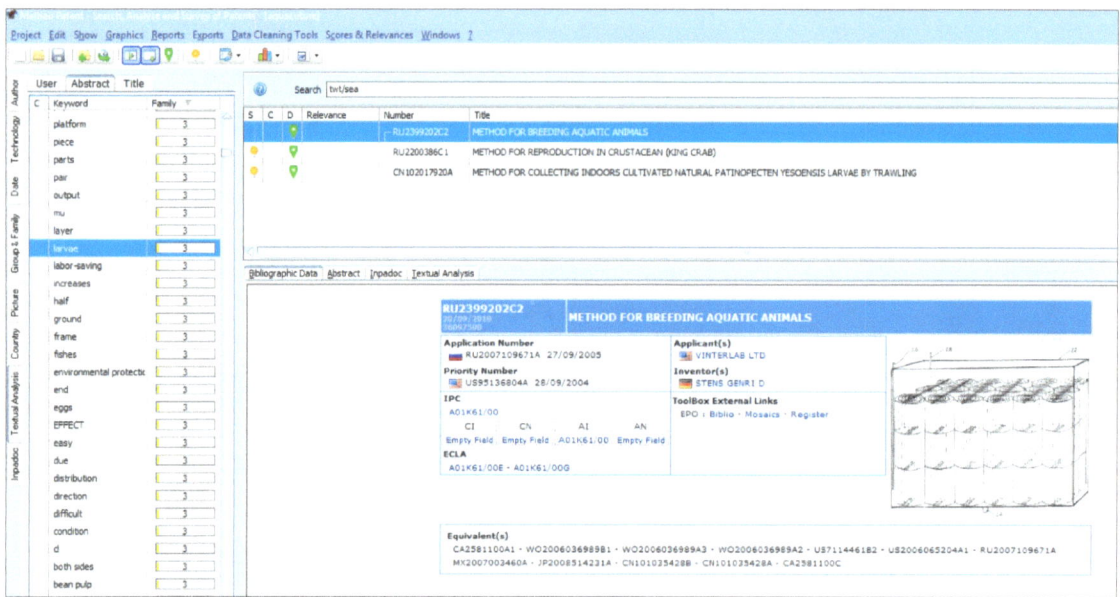

Figure 17 Textual analysis using significant expressions from titles and abstracts

This provides the way to refine the selection or to pick up the relevant vocabulary, or to select the patents covered by various expressions and from them to build up specific group of patents on which deeper analysis may be done. Here we selected the expression larvae, which give three patents the selected one is a US Patent (Priority PR US) extended in Russia, with two authors one US another German.

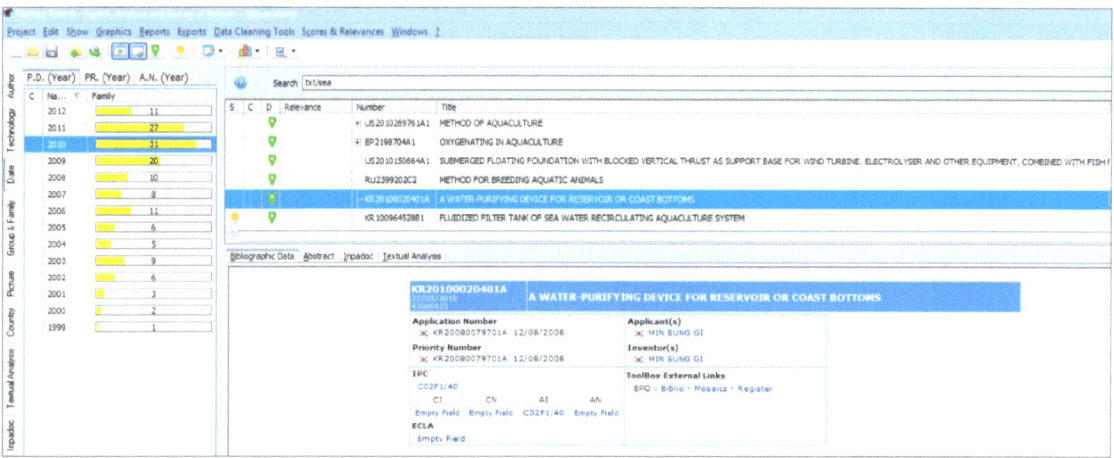

Figure 18 Chronology of the patent dates

Dates chronology shows if there is an interest in the field (growing number of patents); the selection of one date gives the patents related to it, the selection of one patent gives the bibliographic data, the abstract.

From these various treatments above (not a limitative example, other selections being available (IPC 8 digits, inventors, applicants, drawings, etc.), various groups of patents can be build up from the selected patents. These groups will be further analyzed until the moment of the most important patents will be selected for reading. The following figures present various examples.

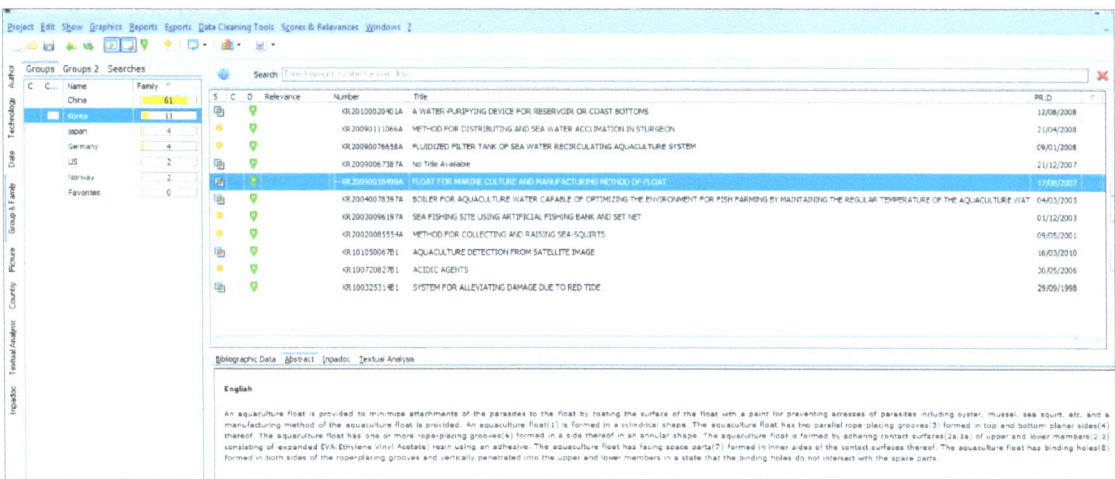

Figure 19 Examples of various groups built up from the former analysis

We built up here various country groups by using the patent selection from priority countries. Note the very large amount of Chinese patents, most of the time not extended out of China and also present as utility model (small patent with a span of 10 years for China). On these various groups further analysis may be done (lists, matrix, networks, etc.) as well as the combination of thee group with other data such as IPC, dates, or other groups (for instance from titles and abstracts significant expression. This this latter combination will conduct to meta strategic matrices.

3.3.3 Specific analysis on the full local database

3.3.3.1 Examples

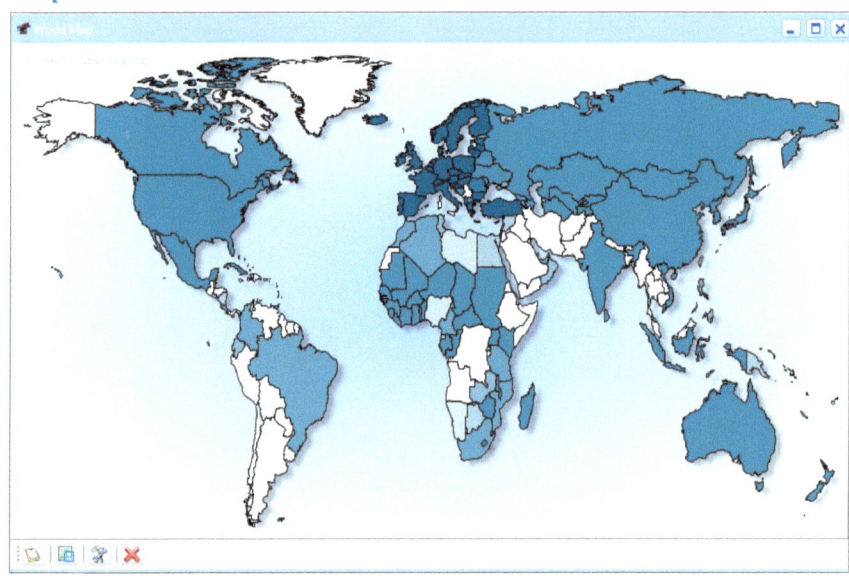

Figure 20 Designated states in the full local database

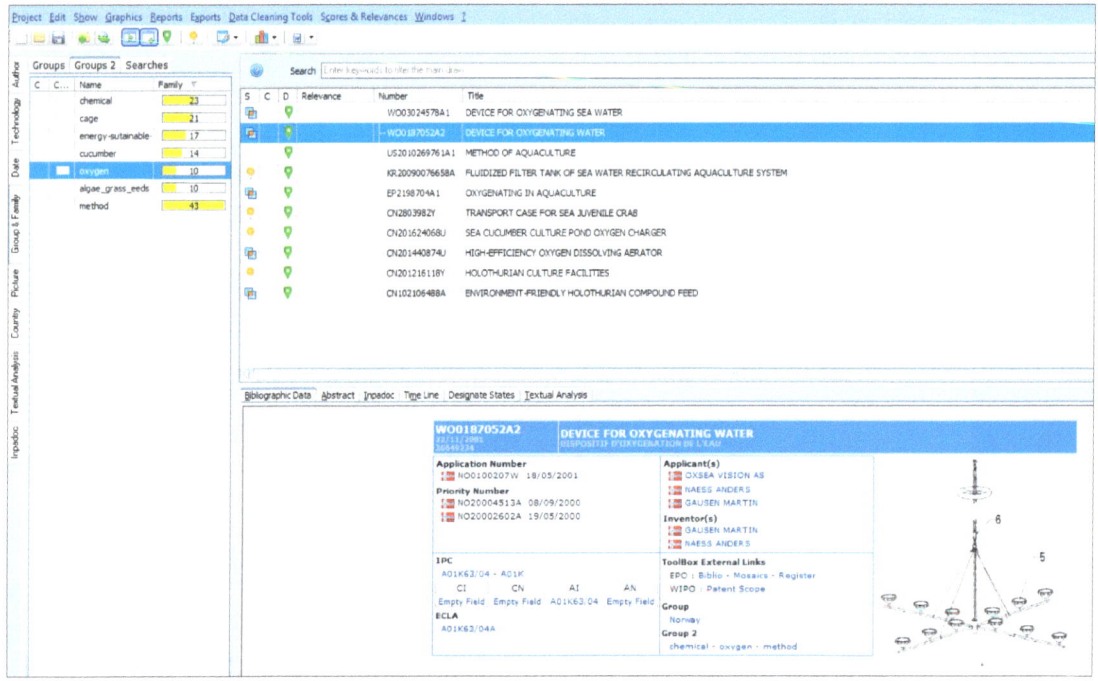

Figure 21 Examples of groups build up from significant expressions from titles or abstracts

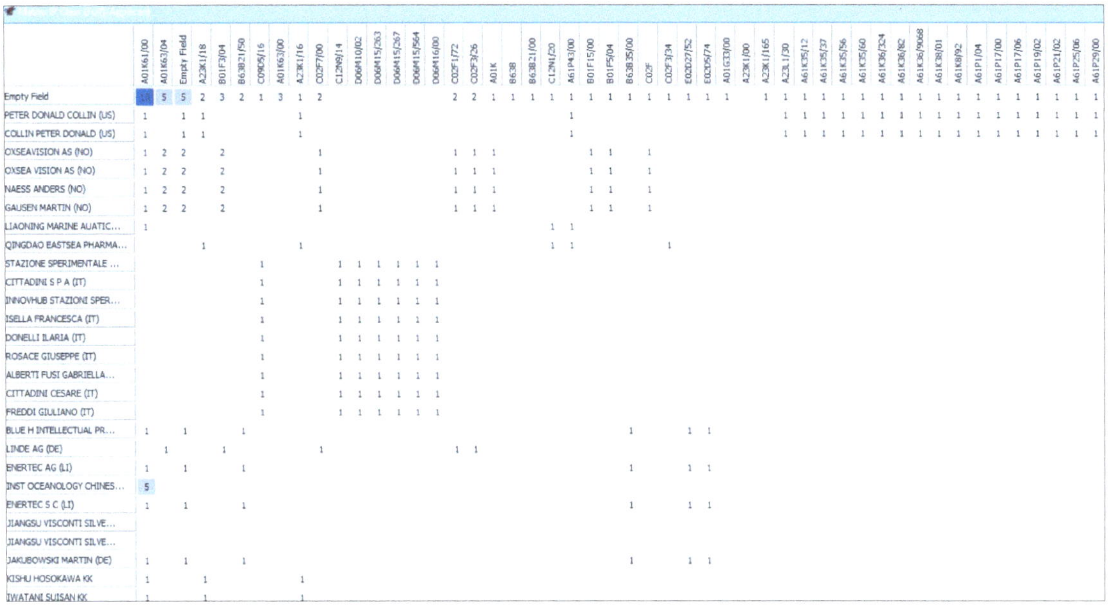

	A01K61/00	A01K63/04	Empty Field	A23K1/18	B01F3/04	B63B21/50	C09D5/16	A01K63/00	A23K1/16	C02F7/00	C12N9/14	D06M10/02	D06M15/263	D06M15/267	D06M15/564	D06M16/00
Empty Field	[illegible]	5	5	2	3	2	1	3	1	2						
PETER DONALD COLLIN (US)	1		1	1					1							
COLLIN PETER DONALD (US)	1		1	1					1							
OXSEAVISION AS (NO)	1	2	2		2					1						
OXSEA VISION AS (NO)	1	2	2		2					1						
NAESS ANDERS (NO)	1	2	2		2					1						
GAUSEN MARTIN (NO)	1	2	2		2					1						
LIAONING MARINE AUATIC...	1															
QINGDAO EASTSEA PHARMA...				1					1							
STAZIONE SPERIMENTALE ...							1				1	1	1	1	1	1
CITTADINI S P A (IT)							1				1	1	1	1	1	1
INNOVHUB STAZIONI SPER...							1				1	1	1	1	1	1
ISELLA FRANCESCA (IT)							1				1	1	1	1	1	1
DONELLI ILARIA (IT)							1				1	1	1	1	1	1
ROSACE GIUSEPPE (IT)							1				1	1	1	1	1	1
ALBERTI FUSI GABRIELLA...							1				1	1	1	1	1	1
CITTADINI CESARE (IT)							1				1	1	1	1	1	1
FREDDI GIULIANO (IT)							1				1	1	1	1	1	1
BLUE H INTELLECTUAL PR...	1		1			1										
LINDE AG (DE)		1			1					1						
ENERTEC AG (LI)	1		1			1										
INST OCEANOLOGY CHINES...	5															
ENERTEC S C (LI)	1		1			1										
JIANGSU VISCONTI SILVE...																
JIANGSU VISCONTI SILVE...																
JAKUBOWSKI MARTIN (DE)	1		1			1										
KISHU HOSOKAWA KK	1			1					1							
IWATANI SUISAN KK	1			1					1							

	C02F1/72	C02F3/26	A01K	B63B	B63B21/00	C12N1/20	A61P43/00	B01F15/00	B01F5/04	B63B35/00	C02F	C02F3/34	E02D27/52	E02D5/74	A01G33/00	A23K1/00	A23K1/165
Empty Field	2	2	1	1	1	1	1	1	1	1	1	1	1	1	1		1
PETER DONALD COLLIN (US)							1										
COLLIN PETER DONALD (US)							1										
OXSEAVISION AS (NO)	1	1	1					1	1		1						
OXSEA VISION AS (NO)	1	1	1					1	1		1						
NAESS ANDERS (NO)	1	1	1					1	1		1						
GAUSEN MARTIN (NO)	1	1	1					1	1		1						
LIAONING MARINE AUATIC...						1	1										
QINGDAO EASTSEA PHARMA...						1	1					1					
STAZIONE SPERIMENTALE ...																	
CITTADINI S P A (IT)																	
INNOVHUB STAZIONI SPER...																	
ISELLA FRANCESCA (IT)																	
DONELLI ILARIA (IT)																	
ROSACE GIUSEPPE (IT)																	
ALBERTI FUSI GABRIELLA...																	
CITTADINI CESARE (IT)																	
FREDDI GIULIANO (IT)																	
BLUE H INTELLECTUAL PR...										1			1	1			
LINDE AG (DE)	1	1															
ENERTEC AG (LI)										1			1	1			
INST OCEANOLOGY CHINES...																	
ENERTEC S C (LI)										1			1	1			
JIANGSU VISCONTI SILVE...																	
JIANGSU VISCONTI SILVE...																	
JAKUBOWSKI MARTIN (DE)										1			1	1			
KISHU HOSOKAWA KK																	
IWATANI SUISAN KK																	

	A23L1/30	A61K35/12	A61K35/37	A61K35/56	A61K35/60	A61K36/324	A61K36/82	A61K36/9068	A61K38/01	A61K8/92	A61P1/04	A61P17/00	A61P17/06	A61P19/02	A61P21/02	A61P25/06	A61P29/00
Empty Field	1	1	1	1	1	1	1	1	1	1	1	1	1	1	1	1	1
PETER DONALD COLLIN (US)	1	1	1	1	1	1	1	1	1	1	1	1	1	1	1	1	1
COLLIN PETER DONALD (US)	1	1	1	1	1	1	1	1	1	1	1	1	1	1	1	1	1
OXSEAVISION AS (NO)																	
OXSEA VISION AS (NO)																	
NAESS ANDERS (NO)																	
GAUSEN MARTIN (NO)																	
LIAONING MARINE AUATIC...																	
QINGDAO EASTSEA PHARMA...																	
STAZIONE SPERIMENTALE ...																	
CITTADINI S P A (IT)																	
INNOVHUB STAZIONI SPER...																	
ISELLA FRANCESCA (IT)																	
DONELLI ILARIA (IT)																	
ROSACE GIUSEPPE (IT)																	
ALBERTI FUSI GABRIELLA...																	
CITTADINI CESARE (IT)																	
FREDDI GIULIANO (IT)																	
BLUE H INTELLECTUAL PR...																	
LINDE AG (DE)																	
ENERTEC AG (LI)																	
INST OCEANOLOGY CHINES...																	
ENERTEC S C (LI)																	
JIANGSU VISCONTI SILVE...																	
JIANGSU VISCONTI SILVE...																	
JAKUBOWSKI MARTIN (DE)																	
KISHU HOSOKAWA KK																	
IWATANI SUISAN KK																	

Figure 22 Automatic benchmarking of Patent Assignee know how. Partial view (full local database)

This is done by building a matrix from the IPC 4 digits and all the applicants present in the local database. But, this matrix may be "corrupted" or "complicated" because in some case the inventor names also appear in the applicant field. To avoid this inconvenience it is necessary to reformat the data or to select the enterprises (real applicants) by building a group of enterprises from the full applicant list. But in this case if there is only the inventor name in the applicant field (personal patent) the patent will not be selected.

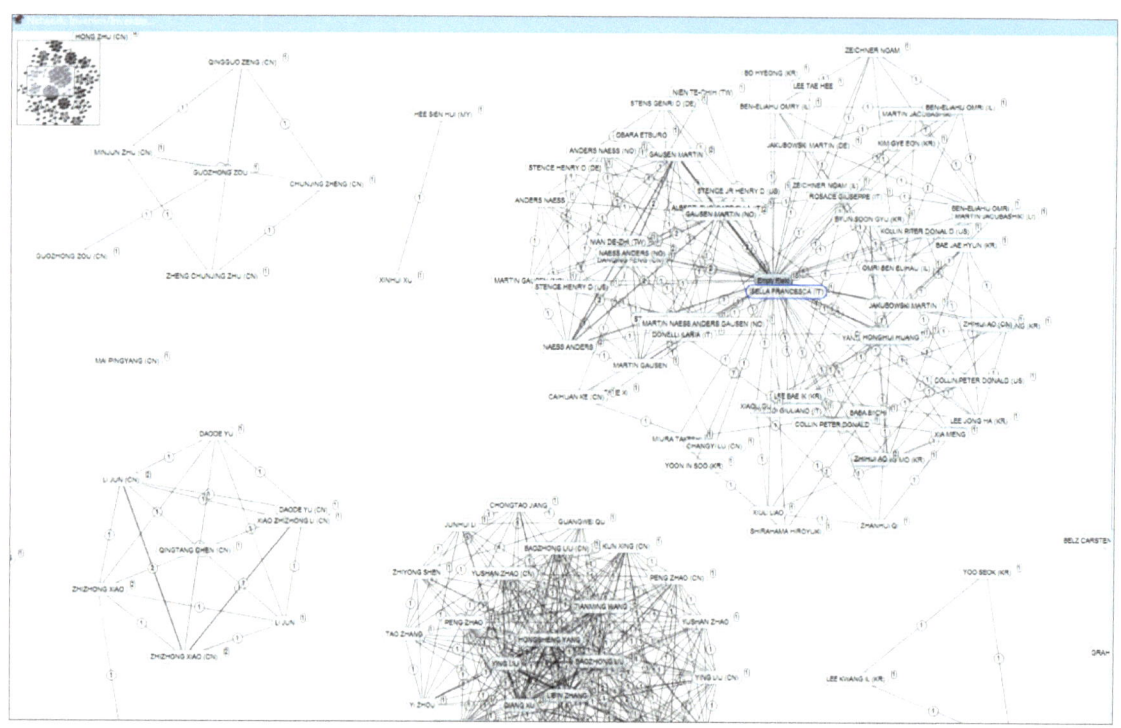

Figure 23 Partial view of inventors networks (full local database)

The network of inventors can be expanded according the need of the user to have a total view of the inventors involved in each networks. The same can be done for applicants to see (if it exists) the network of the multi-applicants (more than one applicant per patent).

3.3.3.2 Specific analysis and strategic matrices

What are the best actors in some of the groups make by the user(s)?

	ZHIHUI AO (CN)	ZHENLE XI	ZEICHNER NOAM (IL)	YUSHE GUO	YUANSHI FENG (CN)	YOON IN SOO (KR)	YEUNG SUI LEUNG (HK)	XU RENDAO (CN)	XINGCHENG TACHUADAO SE…	WINTERLAB LTD (US)	WENG SHAOPING (CN)	VINTERLAB LTD (US)	UNIV XIAMEN (CN)	UNIV SUN YAT SEN (CN)	UNIV SOUTH CHINA TECH…	UNIV HUAIZHONG AGRICOLT…	TIANMA FEED CO LTD (CN)	TIANJIN CNSIC MARINE B…	SUBFLEX LTD (IL)	STENCE HENRY D (DE)	STAZIONE SPERIMENTAL E…	SHIHUA ZHANG (CN)	SHANGHAI HONGBAO GREEN…	SHANGHAI HONGBAO GREEN.	SHANDONG RES INST OF M…	SHANDONG MARINE FISHER…	SHANDONG LIUHE GROUP C…	SEARL SEAFOODS PTY LTD	RYBKOGO RHOZJAJETVA 1 …	ROSACE GIUSEPPE (IT)	REPUBLIC KOREA (KR)	QUANGZHOU NORMAL UNIVER…	PETER DONALD COLLIN (US)	OXSEAVISION AS (NO)	OXSEA VISION AS (NO)	OCEAN UNIV CHINA	OCEAN INST OF THE CHIN…	NORTH GREEN KK	NINGDE DENGYUE AQUATIC…	NIAN DE-ZHE (TW)	NETER NET UTENSIL CO L…	NANKONG AQUATIC PRODUC…	NANHAI AQUATIC PRODUCT…	NAESS ANDERS (NO)	MAI PINGYANG (CN)	LONGZE MARINE PRODUCTS…	LONGZE MARINE PRODUCT …	LONGZE MARICULTURE CO LTD	LIXIAN HJARLIN AQUATIC…	LIU XIANWU (CN)
chemical			1											1	1	1	1	1					1	1	1	1							1	2	2	1			1								2		1	
cucumber	1										1														1	1	1						1													1	3	1		
energy-sutainable-dvt	2	1		1													1	1				1					1					1										1		1				1		
oxygen			1							1			1																			1		2	2						1			2				1		
algae_grass_eeds									1											1	1	1	1				1																	1					1
cage	2	1	1				1						1			1		1						1						1		1			1			1		1	1		1						
method	1				1	1			1	1	1			1	1	1	1	1		1		1	1	1			1	1	1	1		1	2	2	1	1	1	1	1					2					1	1

Figure 24 Partial view. The most relevant company is the company which possess the maximum patents in the maximum of domains.

What is the difference in applied research between USA and Norway ?

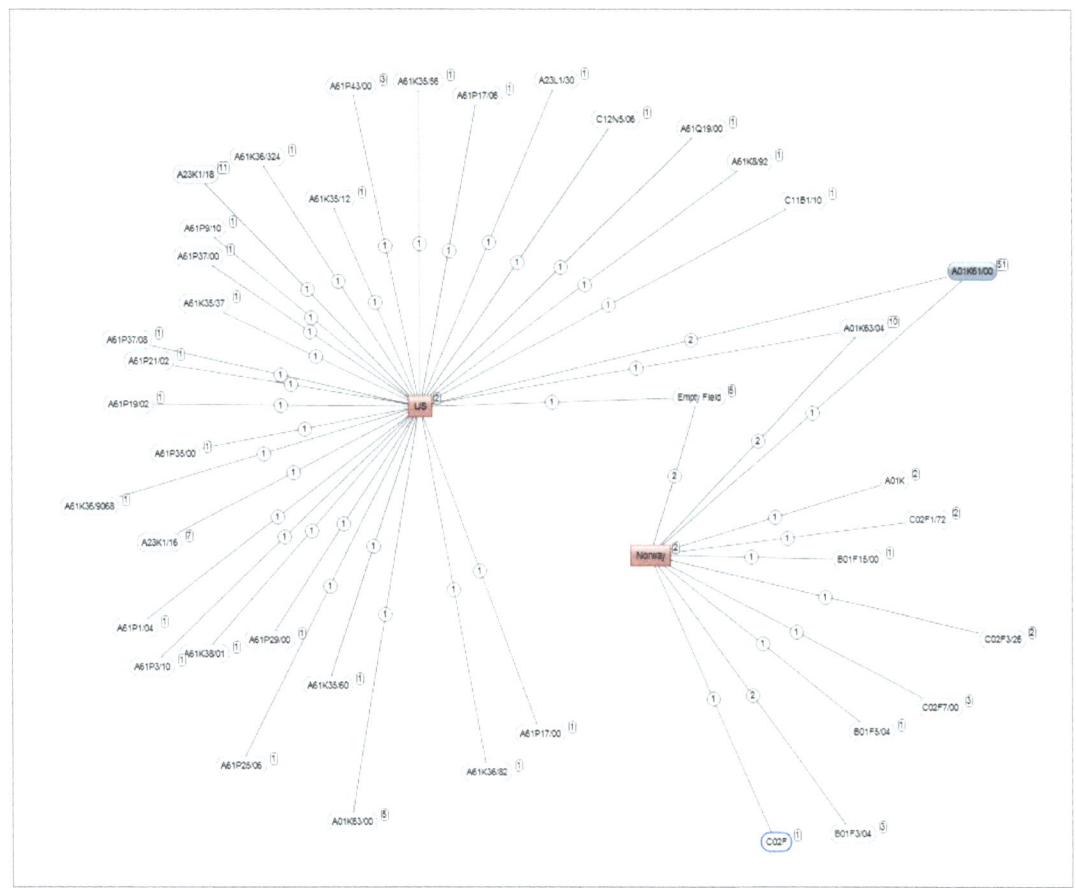

Figure 25 The specific domains are those which have only one link (USA or Norway)

We use it the network build between the IPC 8 digits and the groups USA and Norway. The same can be done for all the groups build up by the user.

What are the application domains covered by the WO and EP patents

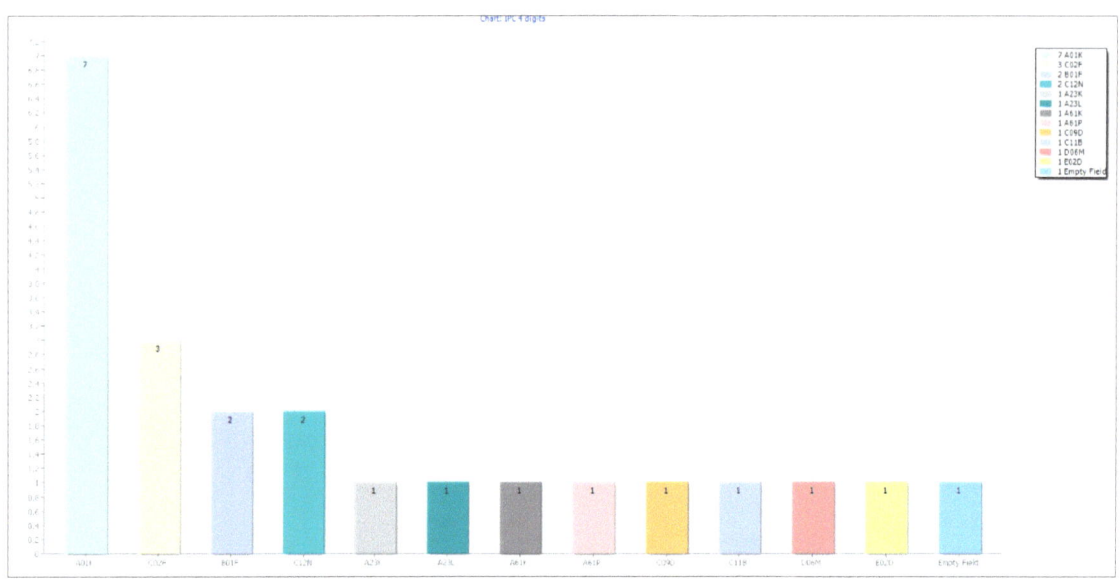

Figure 26 Technologies covered by the world patents (WO)

Priority countries and selected domains

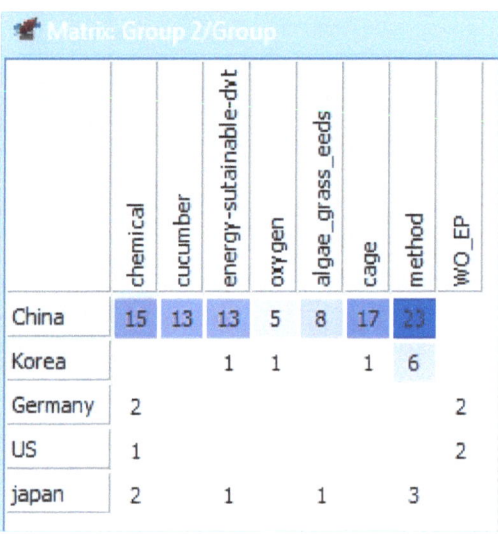

Figure 27 Priority countries and selected domains

This is the result of the combination of the country groups (Priority Countries) with the various selected domains. (It is a meta matrix). The very strong "weight" of China is noticeable but in the group WO_EP the absence of Chine Patents can also be noted.

Universities and research institutes, domain of research

	A01K	B01F	C02F	B63B	E02D	A61K	C12N	A23K	A23L	A61P	C11B	A61D	A01N	C12M	A01G	A23B	A01P	C09D	C12R	A41G	F24H	E02B	G01D	G01K	G01N	G08C	H04M	B01D	B08B	F04D	B65D	Empty Field	A01M	C08L	G06T	A61Q
China	38		9	1		2	4	9	1	2		2	3	2	1	1	2	1	1	1	1	1	1	1	1	1	1	1	1	1	1					
Korea	6		3										1		1																				1	
Germany	4	1	2	1	1																											1				
US	2					1	1	1	1	1	1																					1				1
Japan	4					1		2		1					1																			1	1	
univ_institut	11		2			1	2	3		1			2	2				2	1	1			1	1	1	1	1					1				1

Figure 28 Research domains of universities and institutes

This is a meta matrix. How to read it: This matrix is done by crossing the group of the priority countries (PR=xx) plus the group of the universities and institutes with the IPC 4 digits. All the applications and technologies are on top of the matrix and at the bottom the number of patents granted to institutes and universities. The rows dispatch also the patents numbers by countries of each institute and for each IPC.

Group mapping of the patents granted to universities or institutes

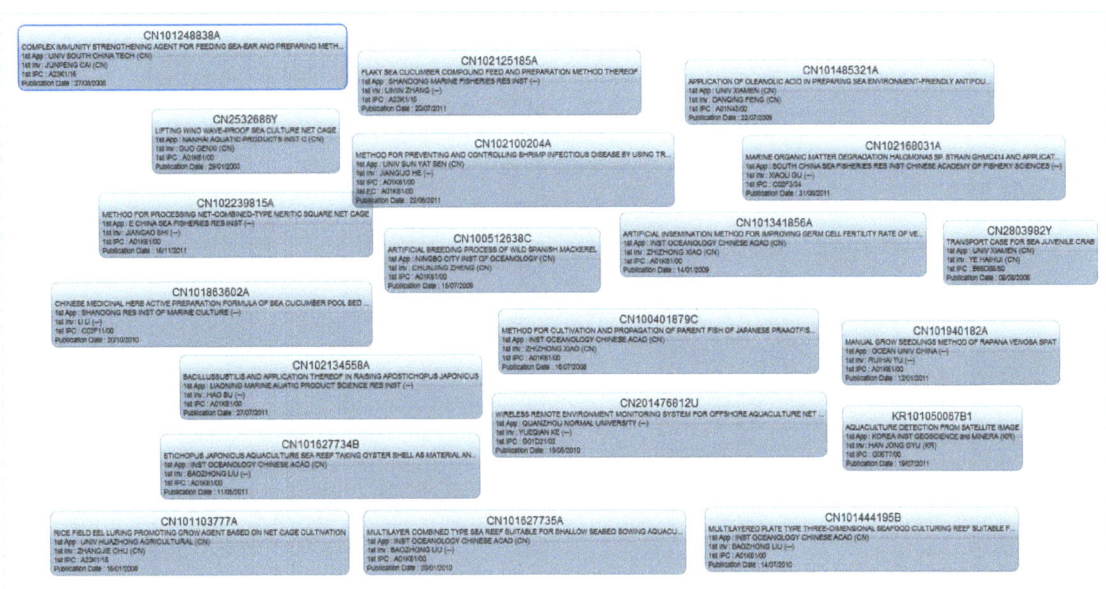

Figure 29 Patents granted to universities or institutes

Detailed analysis of the strategic domains (partiel view)

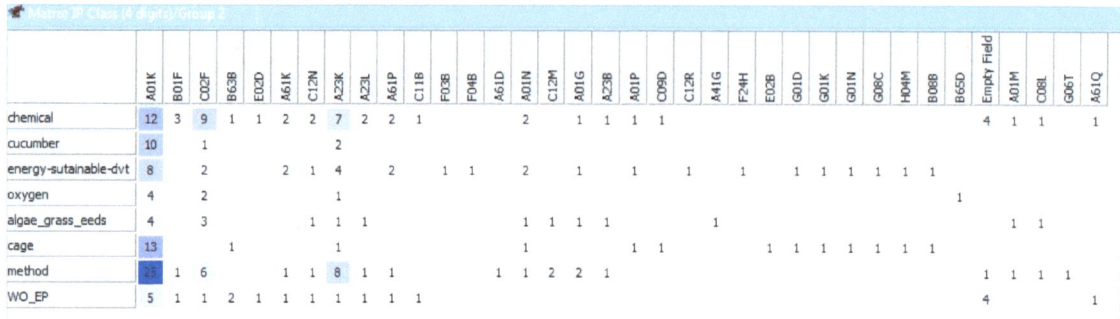

Figure 30 Detailed analysis of the strategic domains

3.4 Data export and other facilities

The local database can globally or partially be exported in different formats (text, XML, CSV, etc .) allows for instance the integration of the data in local intranets (XML) or the export to other software (Matheo-Analyzer) to perform other statistical analysis.

3.4.1 Data Export

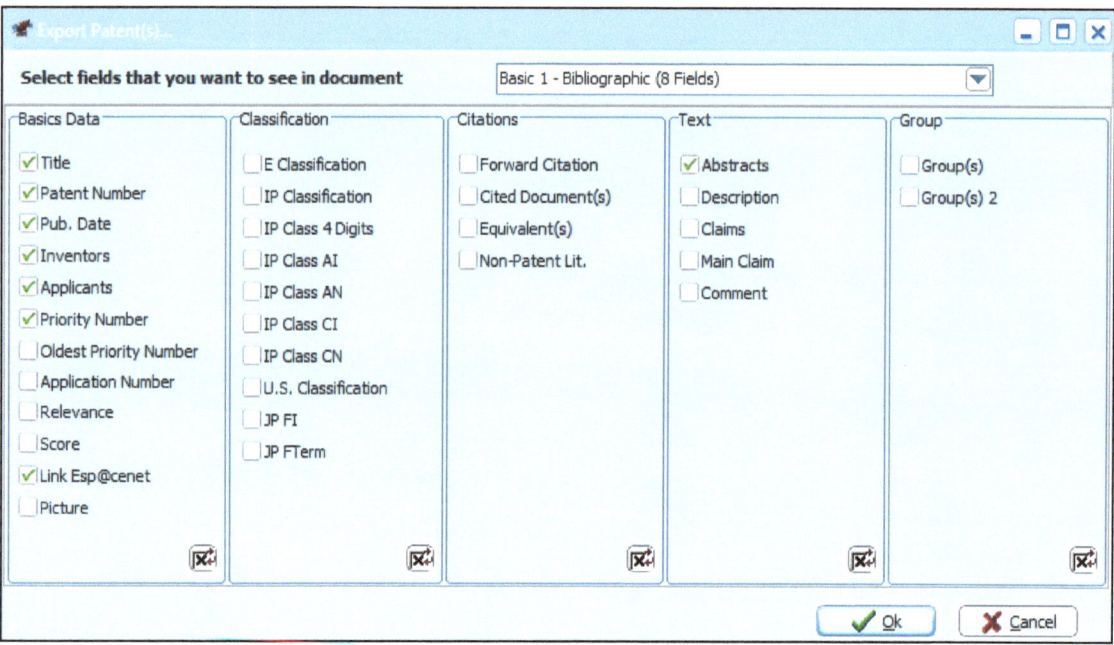

Figure 31 Choice of the various patent fields to export.

Result obtained for the selection of Chinese Utility Models

-1-

TI_EN - TRANSPORT CASE FOR SEA JUVENILE CRAB

PN - CN2803982Y

PD - 09/08/2006

PA - UNIV XIAMEN (CN)

IN - YE HAIHUI (CN); WANG GUIZHONG (CN); LIN QIONGWU (CN); LI SHAOJING (CN)

PR - CN200520007918U 20050328

I4 - B65D

AB_EN - The utility model relates to a transport case for a sea juvenile crab. The utility model relates to a packing device, particularly a packing case which is mainly suitable for transporting sea crab megalops and juvenile crabs from the first stage to the third stage in a multitudinous mode without water. The utility model provides a transport device for sea juvenile crabs, which has the advantages that the utility model can preserve water and moisture, the utility model can reduce the phenomenon that sea juvenile crabs mutually slaughter, the transport survival rate for the juvenile crabs is improved, the structure is simple and the usage is convenient. The utility model is provided with a case body and a cushion layer, wherein at least one vent hole is arranged on the lateral wall of the case body, the cushion layer is arranged at the bottom in the case body and is a soft layer made of water absorption material, the cushion layer is provided with a surface provided with textures, the juvenile crabs can climb and grasp the surface and the juvenile crabs can also adhere on the surface. The material of the case body does not absorb water. The cushion layer has good performance for absorbing water and preserving water, and the moisture in the case can be kept for a long time. At least one vent hole is reserved on the wall of the case, which prevents the juvenile crabs from choking. The texture of cushion material laid at the bottom of the case is soft, which can buffer the bump and the shock during the transportation and can reduce the mechanical damage of the juvenile crabs. The utility model does not need to be filled with oxygen, or be sprayed with water, or be cooled down. The utility model can be used for a plurality of times. The utility model is easily popularized and applied in aquaculture industry.

-2-

TI_EN - BLOOD SEPARATING MACHINE

PN - CN2777988Y

PD - 10/05/2006

PA - SUN FUYI (CN)

IN - SUN FUYI (CN)

PR - CN200520089611U 20050308

I4 - A23K

AB_EN - The utility model relates to a blood separating machine which is suitable for the field of blood products, etc. and can make the wall of a blood corpuscle broken. On the basis of the original tube-type separating machine, a wall breaking device is additionally arranged below a liquid disk of a machine body, wherein the wall breaking device is fixedly connected with a rotary drum through screws and is composed of an annular separating cover, a separating ring, a support pad, a shock insulator and a frustum-shaped separating ring cover, and the inner wall of the separating ring cover is provided with multiple rings of convex teeth of which the longitudinal cross section forms a saw tooth shape or is uniformly distributed with convex stabs. When a separated blood corpuscle passes through the wall breaking device, the wall of the blood corpuscle is broken under the action of the convex teeth or the convex stabs on the inner wall of the separating ring cover. Compared with the prior art, the blood separating machine of the utility model has multiple purposes in one machine, makes separation processing be in a succession state, reduces the investment of devices for manufacture plants, and also provides the convenience for the further processing of the blood corpuscle. The utility model reduces the whole production cost, is beneficial for absorbing organism in the sea by using feed additive processed by the broken blood corpuscle, and promotes the development of marine aquaculture.

Note the presence of the URL which allows automatically to enter in the EPO database to access to a deeper information about this patent (bibliographic description, first page, drawings, full text of the patent).

3.4.2 Data Import

People working with : Delphion (format XML), PatBase (format XML), Micro Patent (format CSV), Derwent (format ISI) can benefit of the Matheo-Patent treatments, Matheo-Patent integrates automatically the format of the former patent management software.

3.4.3 Reports

They are many ways to build up automatic reports. This is done on all the data of local database and various types of pre-defined reports are available.

3.4.3.1 Different types of reports available

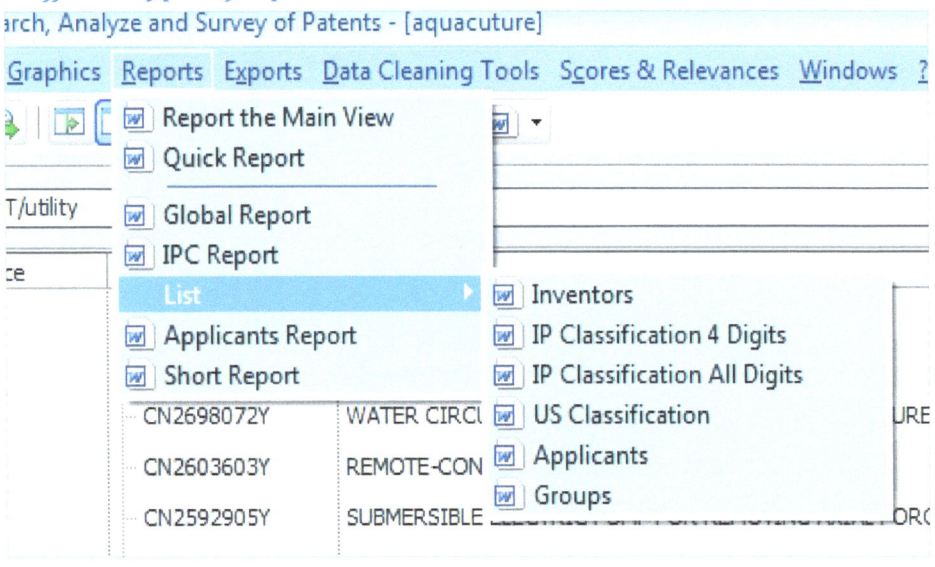

Figure 32 Different types of reports available

3.4.3.2 Short report about the local database

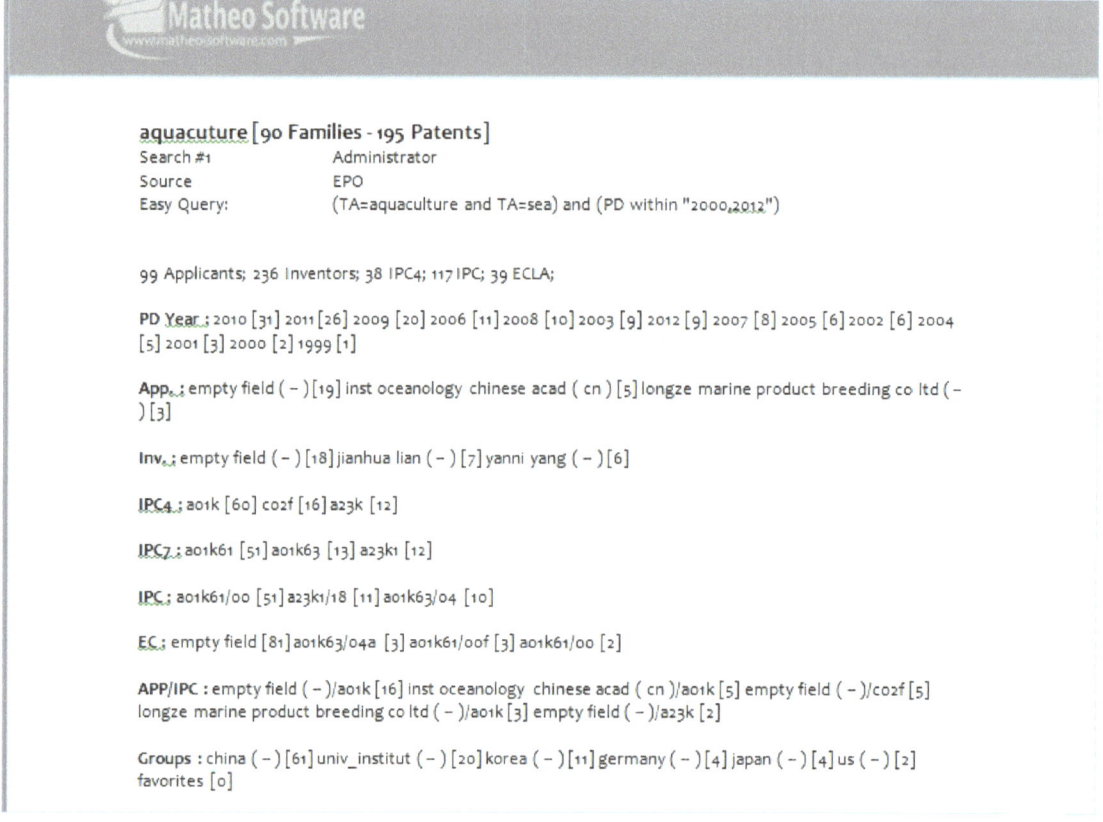

Figure 33 Short report about the local database

3.5 The link between research and technology

We saw in the above paragraphs that it is possible to select patents from the local database. We did it for patents granted to universities or institutes. Here is an extract of the list of the Titles, Applicants and Inventors presents in these patents:

-10-

TI_EN - MANUAL GROW SEEDLINGS METHOD OF RAPANA VENOSA SPAT

PA - OCEAN UNIV CHINA (--)

IN - RUIHAI YU (--); QI LI (--); ZHAOPING WANG (--)

-11-

TI_EN - CHINESE MEDICINAL HERB ACTIVE PREPARATION FORMULA OF SEA CUCUMBER POOL BED MATERIAL MODIFIER AND PRODUCTION PROCESS THEREOF

PA - SHANDONG RES INST OF MARINE CULTURE (--)

IN - LI LI (--); ENFU LIU (--); GUANGBIN LIU (--); ZHAOXING QIU (--); XIAOHONG WANG (--); YAN ZOU (--)

-12-

TI_EN - MULTILAYER COMBINED TYPE SEA REEF SUITABLE FOR SHALLOW SEABED SOWING AQUACULTURE

PA - INST OCEANOLOGY CHINESE ACAD (CN)

IN - BAOZHONG LIU (--); HONGSHENG YANG (--); YING LIU (--); QIANG XU (--); KUN XING (--); LIBIN ZHANG (--); ZONGHE YU (--); QINGYUAN LIU (--)

-13-

TI_EN - STICHOPUS JAPONICUS AQUACULTURE SEA REEF TAKING OYSTER SHELL AS MATERIAL AND AQUACULTURE METHOD THEREOF

PA - INST OCEANOLOGY CHINESE ACAD (CN)

IN - BAOZHONG LIU (--); HONGSHENG YANG (--); YING LIU (--); QIANG XU (--); TIANMING WANG (--); LIBIN ZHANG (--); TAO ZHANG (--); CHONGTAO JIANG (--); YI ZHOU (--); GUANGWEI QU (--); JUNHUI LI (--); ZHIYONG SHEN (--)

-14-

TI_EN - APPLICATION OF OLEANOLIC ACID IN PREPARING SEA ENVIRONMENT-FRIENDLY ANTIFOULING AGENT

PA - UNIV XIAMEN (CN)

IN - DANQING FENG (CN); CAIHUAN KE (CN); CHANGYI LU (CN)

Let us select now the data related to the INST OCEANOLOGY CHINISE ACAD (selected in grey). We get now from these data the title words, the name of the institution and the inventors names.

We may now go to Google Scholar (to use a free academics information database) and see what happens when we use the applicant name and the word CAGES. The result is the following: (partial extract)

Potential eutrophication assessment for Chinese coastal waters [J]

G Weidong, Z Xiaoming, Y Yiping… - Journal of **Oceanography** …, 1998 - en.cnki.com.cn

… 6, Zhu Zhuohong,Xu Meichun and Chen JinsiSouth China Sea **Inst**. of Oceanol. … 9, GUO Fang,HUANG Xiao-ping(LED,South China Sea **Institute** of **Oceanology**,The **Chinese Academy** of Sciences, Guangzhou 510301,China);Advances of Studies on Impacts of Net-**cage** …

Effects of mariculture on coastal ecological environment [J]

T SHU, L Luo, Y WEN - Marine Environmental Science, 2002 - en.cnki.com.cn

... of **Oceanology**, The **Chinese Academy** of Sciences, Guangzhou 510301);A STUDY ON SEAWATER ENVIRONMENT AFFECTED BY **CAGE** MARICULTURE IN DAYA BAY[J];TROPIC **OCEANOLOGY**;1996-02. ... 7, Ji Weidong 11.Third **Institute** of Oceanography,State Oceanic ...

[CITATION] **A STUDY ON SEAWATER ENVIRONMENT AFFECTED BY** CAGE **MARICULTURE IN DAYA BAY [J]**

H Yueqiang, Z Qinghua, W Weiying... - ... **OCEANOLOGY**, 1996 - en.cnki.com.cn

... 2, GUO Fang,HUANG Xiao-ping(LED,South China Sea **Institute** of **Oceanology**,The **Chinese** ... 4, CHEN Ding, ZHENG Ai-rong(Department of Oceanography, Xiamen University, Xiamen, Fujian 361005, China);Contamination of N, P and organic matters from **cage** culture and its ...

Study on Species Composition of Fouling Organisms on Mariculture Cages **[J]**

C Shanmao, Z Congyao, Z Guofan... - JOURNAL OF DALIAN ..., 1998 - en.cnki.com.cn

... Cheng-xing2(1.East China Sea Fisheries Research **Institute**,**Chinese Academy** of Fishery Sciences,Shanghai 200090,China;2.Third **Institute** of Oceanography,State Oceanic ... 6, YAN Tao,LIU Shanshan,CAO Wenhao(South China Sea **Institute** of **Oceanology**,The **Chinese** ...

Study on eutrophication and change of nutrients in the Daya Bay

Y Peng, L SUN, H CHEN, Z WANG - ... BULLETIN-TIANJIN-**CHINESE** ..., 2002 - en.cnki.com.cn

... Key Laboratory of Coastal and Atmospheric Research,Shenzhen-Hong Kong **Institution**,Shenzhen 518057 ... Guojun, Hu Shijin and Zong ZhilunSouth China Sea Fisheries Research **Institute**, Guangxhou 510300 ... 9, Zhu Zhuohong,Xu Meichun and Chen JinsiSouth China Sea **Inst**. ...

This gives the way to more information from the CNKI Chine portal:

Full-Text Search:

CNKI Search

《JOURNAL OF DALIAN FISHERIES UNIVERSITY》   1998-04

Add to Favorite Get Latest Update

Study on Species Composition of Fouling Organisms on Mariculture Cages

Cao Shanmao\\ Zhang Congyao\\ Zhang Guofan\\ Wu Yujing (Department of Aquaculture, DFU)

Species composition of fouling organisms on mariculture cages were studied by suspanding plans and sampling of whole mariculture cages at Dalijia seawater hatchery in Dalian from June, 1995 to June, 1996. 82 species of animals were recorded. The dominant species were Tubularia marina, Caprella scaaura, Caprella equilibra, Hydroides ezoensis, Styela clava, Botryllus schlosseri, Balanus amphite amphite, Mytilus galloprovincialis. Attachment of the fouling organisms occured predominantly during July, Auegst and September. On the net plans monthly maximum biomass of these fouling organisms was 5780.6 g m 2 (wet weight) and monthly maximum surface covered by these fouling organisms was 100%; On the plastic plans monthly maximum biomass of the fouling organisms was 2285.4 g m 2 (wet weight) and maximum covered surface reached 40.3%. The wet weights of the fouling organisms on mariculture cages were higher during September, 1995 to March, 1996 than those during March, 1995 to September, 1995. The fouling organisms on nets of the cages was found to be 5908.1 g m 2 and 90% surface covered, while on the plastic plans in the cages 1958.4 g m 2 and 85% surface covered occured.

【Key Words】: mariculture cages fouling organisms species composition
【Fund】: 大连市科委攻关项目
【CateGory Index】: S944.3
【DOI】: cnki:ISSN:1000-9957.0.1998-04-002

Download(CAJ format)   Download(PDF format)
CAJViewer7.0 supports all the CNKI file formats; AdobeReader only supports the PDF format.

Figure 34 full title of a paper in Google Scholar

Now, let us introduce the full title of this paper in Google Scholar, we obtain (extract):

ECOLOGICAL STUDIES ON THE MARINE FOULING ORGANISMS **AT SOME IMPORTANT PORTS OF CHINA [J]**

LI KIE-MIN, H XIU-MING, LI GUO-ZHEN… - … Et Limnologia Sinica, 1964 - en.cnki.com.cn

… 8, Cao Shanmao\ \ Zhang Congyao\ \ Zhang Guofan\ \ Wu Yujing (Department of Aquaculture, DFU);**Study** on **Species Composition** of **Fouling Organisms** on **Mariculture Cages**[J];JOURNAL OF DALIAN FISHERIES UNIVERSITY;1998-04. …

Where a large amount of references other than Chinese are present. Then from successive queries new vocabulary, names of various academic institutions can be detected. This approach will facilitate the discussion and cooperation with academic laboratories.

Of course there are many other ways such as the use of inventor names (beware of the Chinese names which are transliterated and this may introduce mistakes), or titles words, etc. to be used to query scientific databases.

3.6   Conclusion

All the former results underline how from a precise subject, the APA may show to experts and stakeholders of a peculiar domain a deep insight on what is going on in applied sciences. Because information provided by patents is unique, this information can be used to attract the different stakeholders and to prefigure a cluster development. The use of academic, patents and economic information fills the gap between academic research and industrial concerns and opens the way for the development of PPP Public and Private Partnerships (Academics, industry and political institutions) (Navery 2010).

3.7   Biblography

Dou Henri, (2012) Chinese Strategy in Intellectual Property Management, VSST 2012 Ajaccio, opening conference (available on http://www.ciworldwide.org)

Dou Henri, Léveillé Valérie, Manullang Sri,   Dou Jean-Marie r, Patent Analysis for Competitive Technical Intelligence and Innovative Thinking, Data Science Journal (DSJ), Vol. 4 (2005) pp.209-236

Dou Henri , Benchmarking R&D and companies through patent analysis using free databases and special software: a tool to improve innovative thinking, World Patent Information, Volume 26, Issue 4 , December 2004, Pages 297-309

Erikson Per, (2006), Strategic Intelligence and Innovative Clusters − A Regional Policy Blueprint Highlighting the use of Strategic Intelligence in Cluster policy. Interreg III C (European Community) Centro Formativo Privinciale, Guiseppe Zanardelli,   Azienda speciale de la provincia de Brescia, Interreg III C, VINNOVA, Brics-workshop - Aalborg Swedish Governmental Agency for Innovation Systems, 13th Feb 2006

Guellec Dominique,, Bruno van Pottelsberghe de la Potterie, (2001), The internationalisation of technology analysed with patent data, n° 30, pp.1253–1266

http://gb.espacenet.com/search97cgi/s97_cgi.exe?Action=FormGen&Template=gb/EN/home.hts

http://www.matheo-software.com a trial version is available free of charge

IPC (2012), http://www.wipo.org/classifications/fulltext/new_IPC/index.htm6

Inpadoc (2012) Present the legal status of patents;
http://www.epo.org/searching/subscription/raw/product-14-11_fr.html

Navery Nicholas, (2010), Public and private Partnerships, second edition, Editor Lavoisier
S.A.S.

Rostaing Hervé, (1996), La bibliométrie et ses techniques, Sciences de la Société,
Collection "Outils et Méthode

WIPO (2012), http://www.wipo.int/portal/index.html.en

Yanhong Liang , Tan Runhua,(2007)   A test mining-based Patent Analysis in product
innovation process, in Trend in Computer aided Innovation, ed. Noël Lean-Riva, Spinger
IFIP, p. 89

Zoltan J and David B Andrestch,(1998), Innovation in large and small firms. An empirical
study, The American Economic Review, vol.78, n°4, pp. 678-690

4   Automatic Patent Analysis catalyst to develop R&D innovation

Henri Dou

Director of Atelis (Strategic Work Room of the ESCEM – France Business School),

University Professor

douhenri@yahoo.fr   http://www.ciworldwide.org   http://www.atelis.org

Abstract: Innovation is one of the ultimate ways to develop and maintain the competitive advantages of companies and research laboratories. If many research works have been published in this area, there is nevertheless a lack of practical information. In this paper we will present why APA is so important to catalyze innovation and to open a large window on the use of the enterprises and laboratories competencies. After the presentation of the mechanism of innovation, the necessity to develop Public and Private Partnerships (PPP) will be underlined. But, to achieve such a goal, the pre-clusterization of the stakeholders of an area of development is a crucial step. APA again will be of a great help to aggregate people which have the same concern in a given area. APA will also play a great role, because patents are a link between research and applications. It is noticeable than in research papers, the patents are quasi never cited and yet, patents contains author names, applicants, words in titles and abstracts which can be used to retrieve fundamental papers. The language of patents and the drawings are generally understood by both enterprise people and researchers and it facilitates the dialogue between both. The practical aspect of the APA will be presented by using the Matheo-Patent facilities.

4.1 The mechanism of Innovation

The presentation of the innovation in two major steps is due to the work done in the InteRegIII program sponsored by the European Union, presented in different papers, the report by Vinnova for some of these results as well as the work of Canadian researchers. We are not speaking here of innovation groups or other systems which can be set up in companies or laboratories, giving the way to roadmapping, etc. What is emphasized here is the global mechanism which underlines what research and industry must now work together.

4.1.1 Innovation

It is easy to understand that:

- The Government gives money to the universities or to research institutions to help them to build up competencies and knowledge.
- But it is also understandable, that these competencies and knowledge must be used to develop products and services to reach the export markets.

These two steps are fundamental. If the first one is often taken for granted (especially in Latin countries) this is only part of the system. Since, if the competencies are not used, they will become obsolete, and the payoff of the money spent in research will be very low. This is the reason why people are speaking now of a research "economy driven". This is also what the researchers may think to give back to the society part of its investment. This move is called the SRR Social Research Responsibility and is now a growing concept especially with the persistent crisis.

The work of Michael Porter, of the Dutch school, bring to the scene that the best conditions to innovate are at the intersection of the industry – research – and regional or national political institutions. This is called clusters or in some countries poles of competitiveness and it is related to the triple helix research.

4.1.2 The incentive

To develop a cluster, it is necessary to begin to show to the research sector and the enterprises that opening a dialog between both will be interesting. To get the system working, it is necessary to get a strong incentive. A too rigid order or direction to follow, if it is imposed from the top will not work. This is the reason why, one of the requirements to succeed in clusterization is to attract the stakeholders by "something" which will be valuable for the laboratories and for the enterprises.

The best way is to show to both, that with their knowledge and competencies they will be able to develop products and services to make various proposals and to get money. This is the type of drive which helps to keep the dialog running. We believe that this is not the scientific publications which will attract the SMEs, since they often have no people facilities to understand the whereabouts of research. But they understand patents, competitor names and technology areas. In the same way the laboratories will understand what can be done with their competencies if they are able to express them in terms possibly used in patents.

4.1.3 The pre-clusterization

To start it, the mapping of the competencies and knowledge can be done for research and industry of the same sector and especially in the same region[10]. This aspect is more or less political since an impulsion is necessary to start the system. After can be developed a global methodology based on patents analysis. The mapping of the competencies and knowledge will allow the Competitive Intelligence unit to develop its strategy and to detect the information which will be accepted by both and will feed the discussion. For the readers interested in the prerequisites necessary to start a cluster, this information is well described by different South Korean and Thai reports. But for the following part of this paper, we will focus on APA and its the expected results.

4.2 The patent databases and the International Patent Classification

As we said above, the patents databases are now a must for all the people who want to develop University (Research)-Industry cooperation. Patents, opens a large window on applications and products development (Yanhong 2007), (Guellec 2001), (Zoltan 1998). They can be used in this case, not to protect an invention, but to provide a way to underline what others from university, industry, research centers, individuals did with knowledge and competencies very close from their own. It is also interesting to see that what is published in patents is most of the time not published elsewhere and also that

---

[10] We will not develop here the way t create nex added value products from various companies which have complementary knowledge. In this case patent analysis can be useful at the end, when specialists and experts will have by brain storming, deduction, and expertise determine what can of products or services can be develop from this potential cluster.

patents are practically never cited in academics bibliography. Then the patent analysis dealing with a specific field prefigures a virtual cluster which, for the local stakeholders will prefigure also what can be locally developed. Various aspects of the use of text mining or APA or biblbiometrics (Rostaing 1996) provide examples of the potential aggregating power of patent information.

4.2.1　The patent databases available

The choice of the patent database to help people to "see" what his done with their competencies and knowledge is relatively simple:

- The EPO (European Patent Office) makes available free the world patent database which covers more than 90 countries and contains more than 65 million patent notices.
- The US Patent Office makes also available for free two databases which covers the US patents granted and on demand
- Google Patent which allows the access of the US patents in full text

For the national offices, they have their own databases, but often they are not available online free, or they are not very accurate (for some of them). We can say that most of the patents, even Chinese are available through the EPO World Patent Database. This database is of primary importance, because in automatic patent analysis, the goal is not to search a single patent or a very limited number of patents which respond to a very precise query. Sometimes this is necessary, but to look widely at the interface of a subject it is necessary to perform a large query which will give rise to a large number of patents. This is the reason why the APA (Automatic Patent Analysis) is necessary to determine what is going on in this subject. Most of the time we will use the EPO World Patent Database since it is necessary to look on a world large base. We checked also the SIPO database (Chinese Patent Office) and the results obtained were the same than the one obtained with the EPO World Patent Database.

4.2.2　The International Patent Classification (IPC)

This classification is important since the patents are not described by key-words. Only the words present in the titles and abstracts are indexed (except for Google Patent). Then to differentiate the applications, products and technologies a classification has been built. This classification (the IPC) is present in all patents even if there are other classifications such as the US classification, the Japanese classification, the European classification. The IPC is made of a series of letters and figures, up to 8 digits. The more digits the most precise is the description of the class. Are three main ways to access the classification, all of them are available via Internet and various patent sites (EPO, WIPO for instance).

4.2.2.1 *You have one class: example A61K*
You go to the site of the EPO, and look for the part dealing with the IPC. Perform a query with A61K and you will get the corresponding class and under-classes

4.2.2.2 *You search for the IPC(s) corresponding to a particular subject*
Go to the site of the EPO in the part dealing with the IPC and perform a query with the words describing your subject. The database will return the corresponding class(es) and under class(es)

4.2.2.3 *Using the catch words*
The catch words are a list of words which described various parts of technologies, applications and products and which are associated with the different IPC covering the subject. Go to the site dealing with the catch words presents on the WIPO (World International Patent Organization) and perform an interrogation of the catch word list (about 20.000 terms). If the word(s) that you use for the query are present in the list you will get different classes related to this word(s). You will make after you own choice.

4.3   What can be done with APA
To perform APA we generally use the Matheo-Patent software permanent software presents on your computer and which allows to query and download patents from the EPO World Patent Database and the USPTO Patent Databases.

4.3.1   How Matheo-Patent works
This software and various applications have been described in different papers among which the following can be used: (DOU), (DOU), (CIW). The figure 1 indicates the different steps of this process. The data obtained after the query of the available databases (World Patent, USPTO) allow the selection of various numbers of patents which can be downloaded. The software creates locally a formatted database, which will allow an immediate bibliometrics treatment (lists, matrix, networks, reports, etc.). This local database can be updated if necessary. The bibliometrics treatments of the local database created by Matheo-Patent cannot be done manually since the number of data

to analyze is too large. This is why when using patent analysis you will create new information this former being unique since not available in another way.

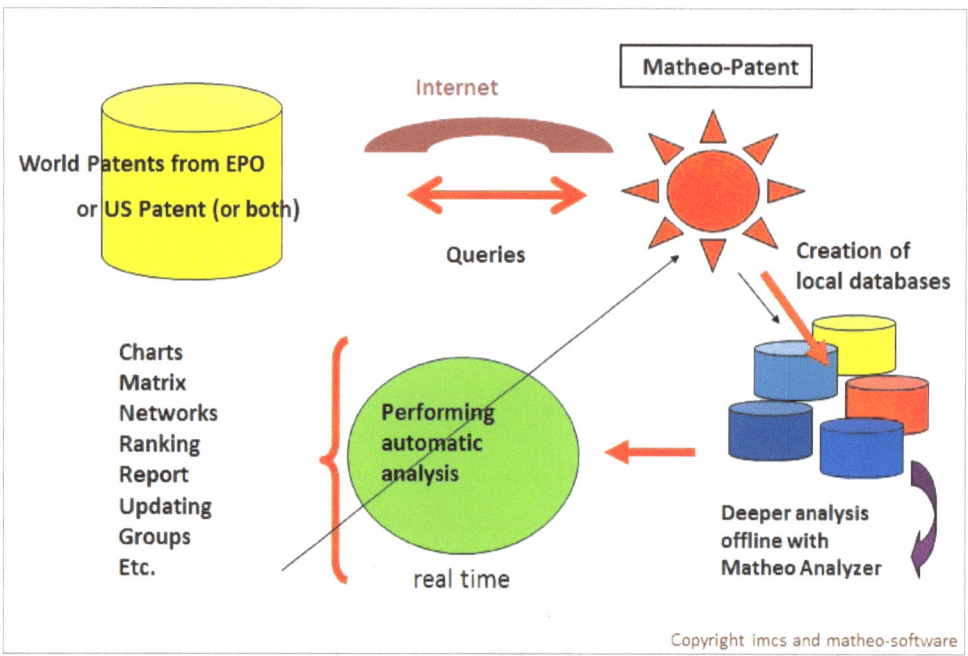

Figure 35 How Matheo-Patent works

4.3.2   What can be done with Matheo-Patent

Most people dealing with the strategy of a company to its future to the trend of technologies for instance will ask various questions such as the ones related to the SWOT analysis, to the Porter's Diamond etc. The patent information when it is analyzed on a large scale will provide many answers related to the above questions:

- Who is doing what? The answer is given by the analysis of the applicant field and by building the matrix applicants/IPC (4 or 8 digits), if necessary groups of applicants can be done to have a most precise matrix (automatic bench marking)
- Where? The analysis of the Priority country will provide the information.
- With whom? The network of applicants will show all the links created through the presence of different applicants in the applicant field of the patents
- What are the groups of inventors working together? The network of the inventor will provide the result? As for applicants, groups of inventors may be selected to build more accurate networks. The same can be done with the network

57

applicants/inventors. This will provide the potential of authors involved in the work of the different applicants.

- What are the competencies of the inventors or of a group of inventors? The matrix of the IPC/inventors will be the solution.
- What is the trend in the development of a technology? The Matrix of the publication dates versus the IPC is also the solution.
- What are the possible new entrants in one field? The matrix of the application date and the applicants will provide the response.
- How can I follow in time my query? Matheo-Patent provides the way to update your local database using the same query, or to implement your local database by using a different query, or both if you like.
- What are the patents which are directly linked to my preoccupation(s)? The use of the significant expressions present in the titles or the abstracts will help to build up strategic patent groups and to analyze them in details as above.
- How can I build up part of my report issue from the APA? Matheo-Patent provides various pre-determined reports that you will have only to implement with your own analysis.
- How can I transfer the selected patents to a Web site or a co-operative platform? Matheo-Patent provides various export facilities (Text Format, CSV, XML)
- What happens if the Applicants and inventor names have different spelling? There is a tool to clean up those fields and reformat the data.

4.4   Example dealing with WELDING

Welding is an operation which is often performed in the development of various machines components, cars, etc. and the technologies which are used in welding change when improvements are made on the various apparatus used for to weld various material or when the technologies of welding changes.

In the following part of this chapter we will show how patent information analysis may help to facilitate the development of various strategies

The figure 1 shows the result of the query of the EPO (Worldpatent database) with:

Welding AND apparatus   selected in the patent titles

It will be also possible to run the same query on titles and abstracts words, but in this case the amount of patents retrieved will be too large. Out of this query we have built up a local database with the 266 patents which have for patent date 2012. Note the large amount of patents retrieved, 1541 from an interval of dates from 2009 to 2012. This really indicates that in the field of the development of new welding machines or apparatus there is a lot of activity in the world. Here, we performed a query which is centered on the welding development and most especially on the apparatus and technologies linked to them. If we were interested by specific part of welding we should have built up a query with other words or even a combination of words and IPC. The choice to perform the query on the titles or the abstracts words depends on the user. Titles words indicate generally that the patent is really centered on the subject. The presence of the same word in the abstracts indicates that part of the patent may deal with the subject. We suggest that depending of what you may search you extend or restrict the number of patents to analyze.

Example of the results obtained when the catch words list is queried with the word "welding":

WELDING B23K
> burners for WELDING F23D
> joining garment parts or blanks by WELDING A41H 43/04
> making tubes with welded seams B21C 37/08
> manufacture of WELDING -wire or of WELDING -rods B23K 35/40
> screens or hoods for WELDING A61F 9/00, F16P 1/06
> WELDING by heat or pressure in box or bag making B31B, B68F
> WELDING of electric connections H01R
> WELDING of metal B23K, F16B 5/08, F16B 11/00
> WELDING of metal in the manufacture of chains or chain links B21L 3/00, B21L 7/00
> WELDING of plastics or substances in a plastic state in general B29C 65/02, B29C 65/72, B29C 65/74
> WELDING of wear-resistant parts on chain links B21L 9/08
> WELDING of wires together B21F 15/08, B21F 27/10

According to your subject the combination of worlds and IPC can be very efficient.

In this case, the year 2012 was chose only for the purpose of demonstration.

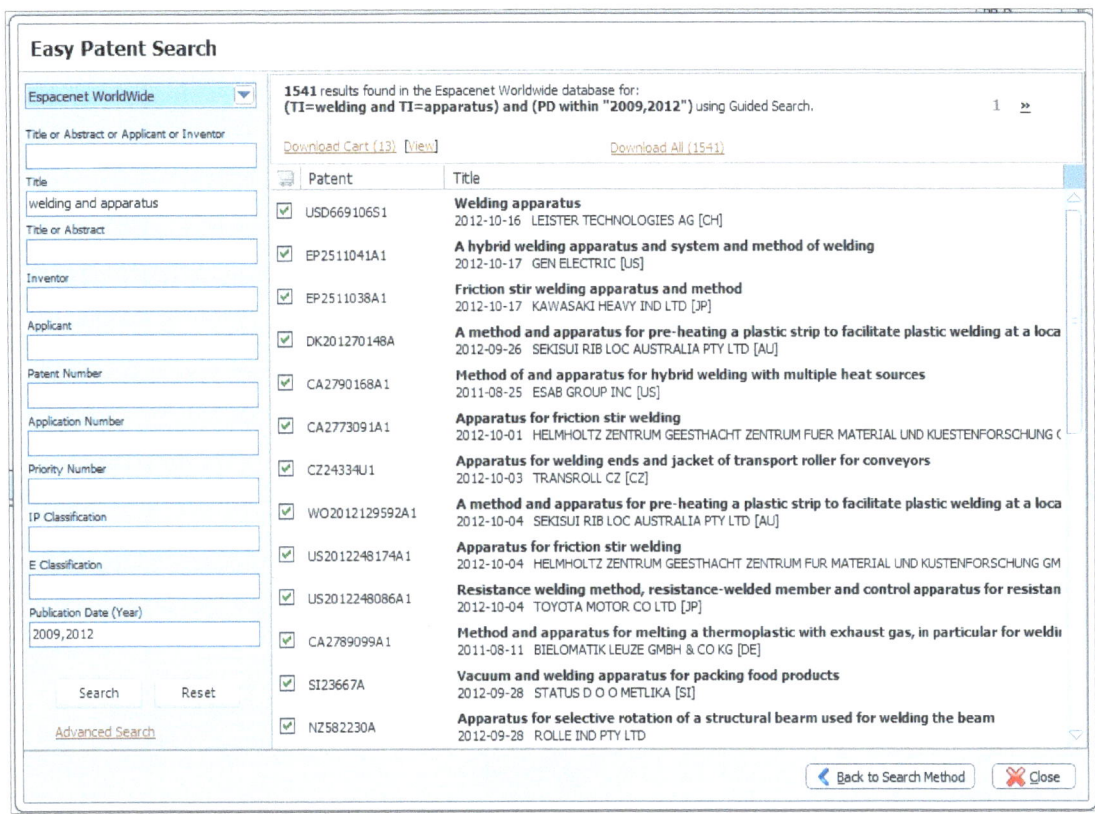

Figure 36 Result of the query welding and apparatus in titles from the EPO worldpatent database

4.4.1 Basic analysis

The figure indicates how to analyze the technologies concerned by the various welding apparatus described in the patents.

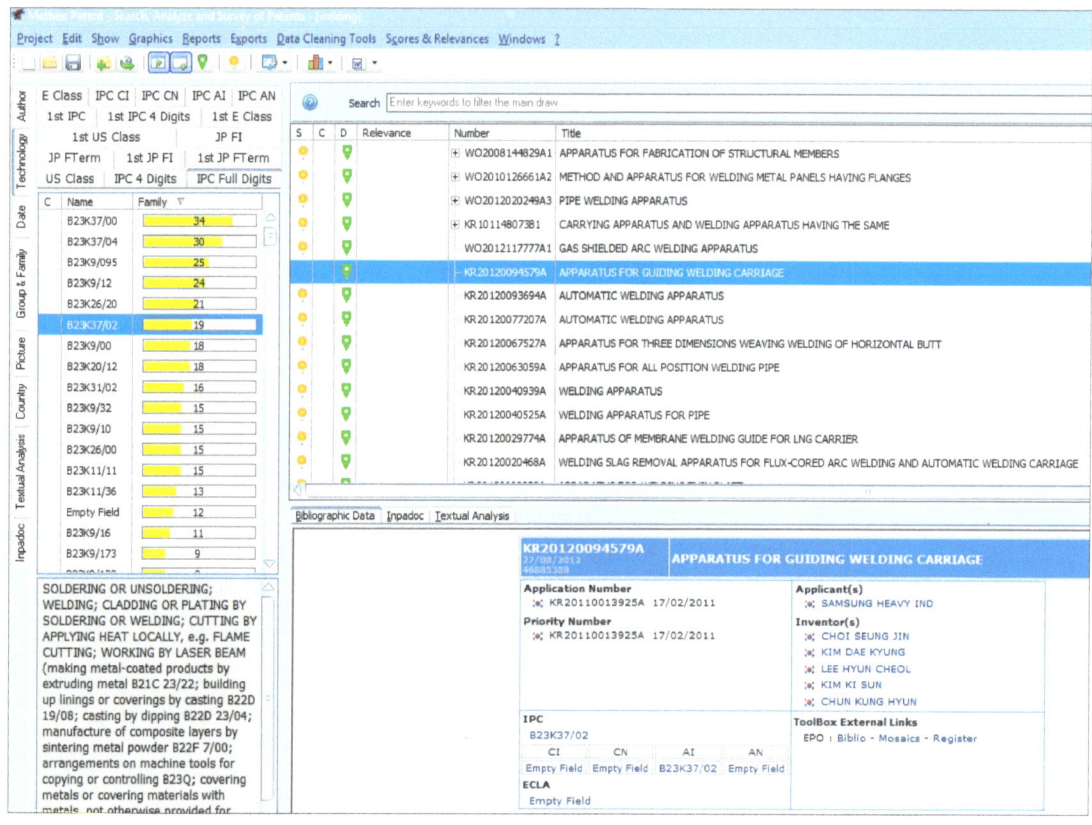

Figure 37 he various IPC 8 digits covered by the selected patents

The figure indicates how to move from the nomenclature of and IPC to its meaning

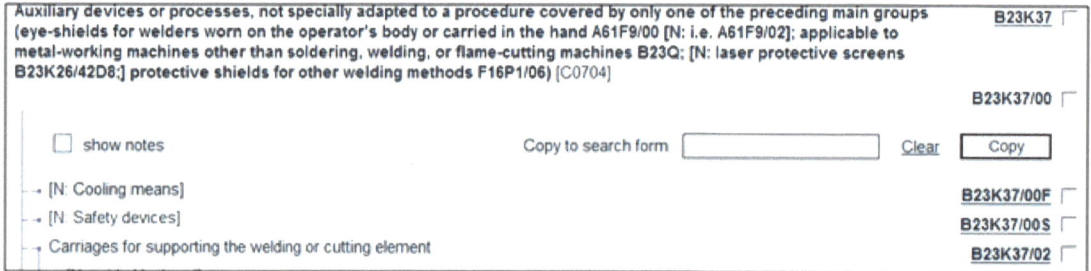

Figure 38 From the IPC nomenclature to its meaning

The figure indicates how to select patents from significant expressions present in the titles and abstracts of the patents present in the local database.

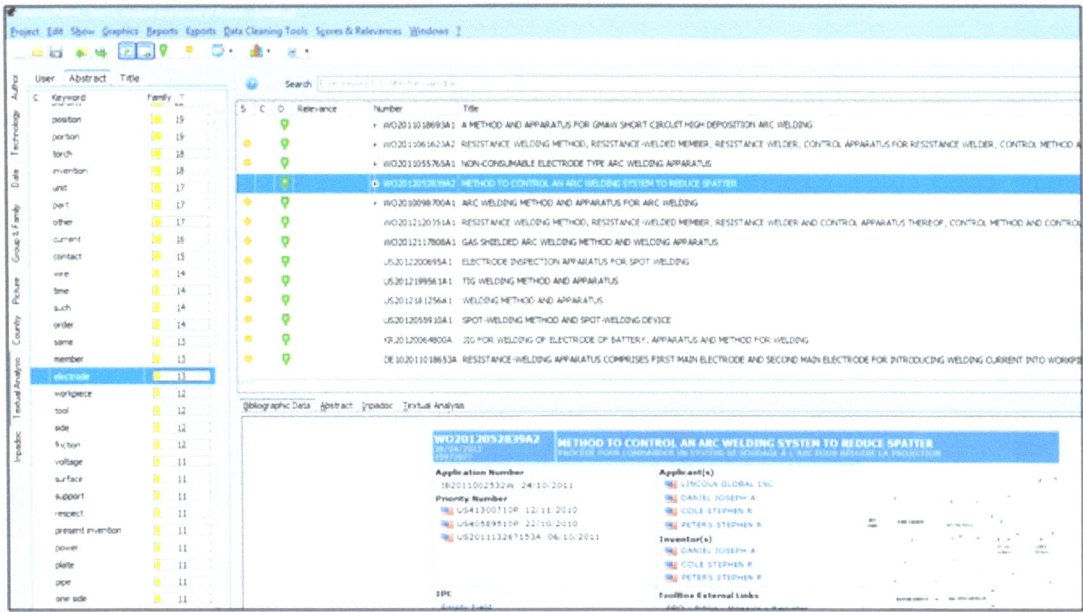

Figure 39 Selection of patents from significant expressions from titles and abstracts

In the same way, we can work on dates on applicants, inventors, and even drawings, which for certain technologies are important. A local database of the drawings present in the selected patents is done, providing a new type of choice, not by using words or codes, but using a direct visual choice.

4.4.2 Building strategic groups

From the above method of selection we will build two groups, one called "voltage" and the other one "welding systems":

Group "voltage": electric resistance, electrode power, electrode, voltage, power

Group "welding systems": laser beam, arc, laser welding, arc welding method, traditional friction, ultrasonic welding, friction welding

From these two groups, we can now build up lists and combinations which will help to have a better understanding of what is going on in these two strategic domains.

4.4.3 Examples of various analyses

The figures underneath represent some of the correlations which can be done. The figure 5 is the automatic benchmarking, using IPC 4 digits of the applicants (companies) involved in the group voltage.

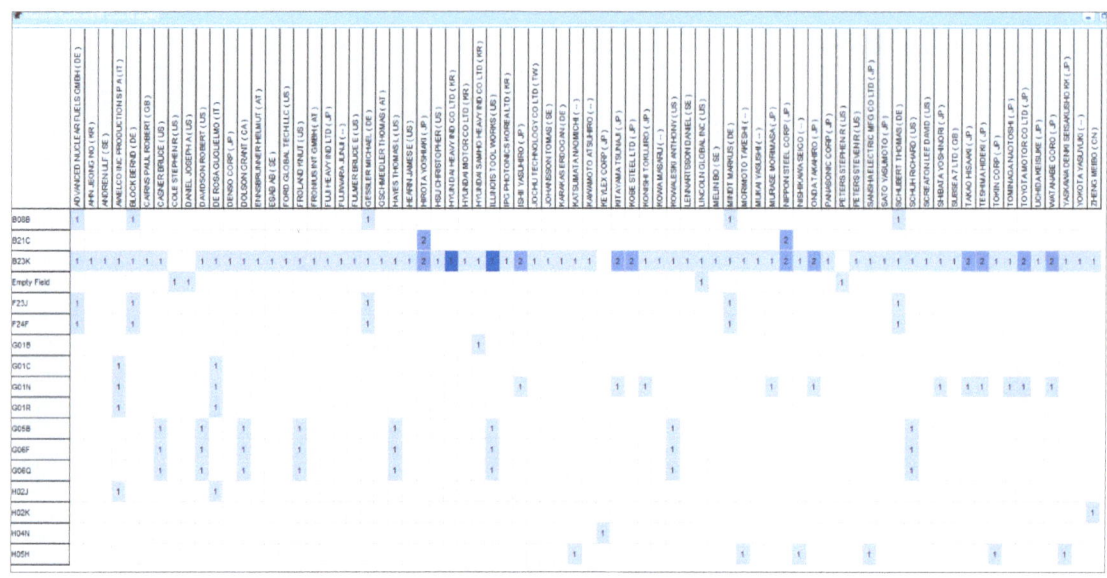

The figure represents the networks of applicants working in the fields of the group "welding systems". This graph shows the co-operation between companies. That is to say, the companies which are present in the same time in the same patent and then which gets a link together (propagation network). It is also necessary for the interpretation of the network to realize that sometimes the inventor names appear in the applicant field. This depends upon the way that the patent has been written and also of the practices in the applicant company.

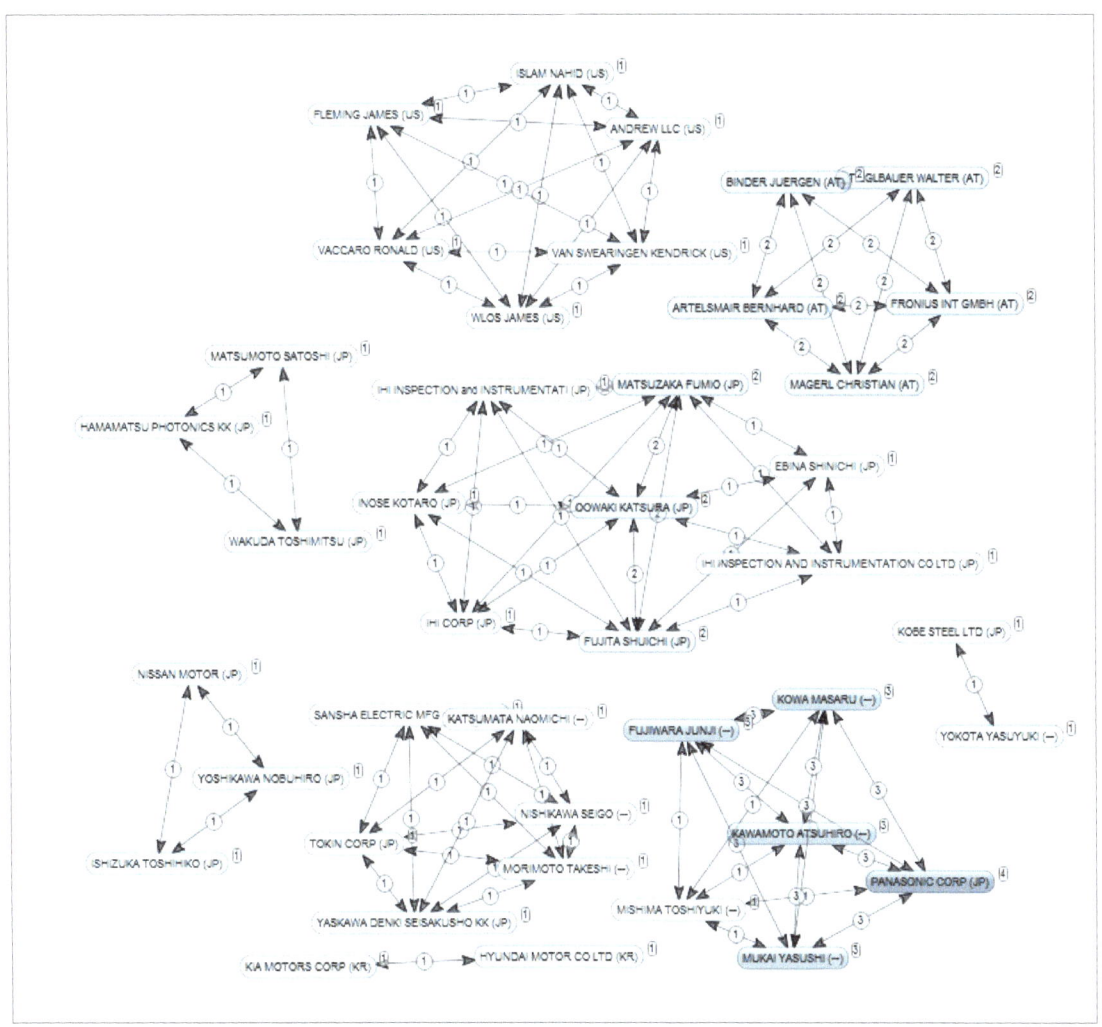

Figure 41 Applicant networks in the field of welding systems

The figure is an example of a meta-matrix. It represents which companies are working in the two fields and those which are common in both. The companies present simultaneously in the two fields should be analyzed first as potentially most important.

Figure 42 Companies present in the two groups "voltage" and "welding systems"

64

The figure represents the countries engaged in the development of technologies and apparatus for welding in the two groups "voltage" and "welding systems". We first build up a group for each main country (number of patents having with the priority in these countries) present in the local database and after we make the network of these groups with the groups "voltage" and "welding system". The result is a meta-network which represents the degree of involvement of the different countries in the two groups.

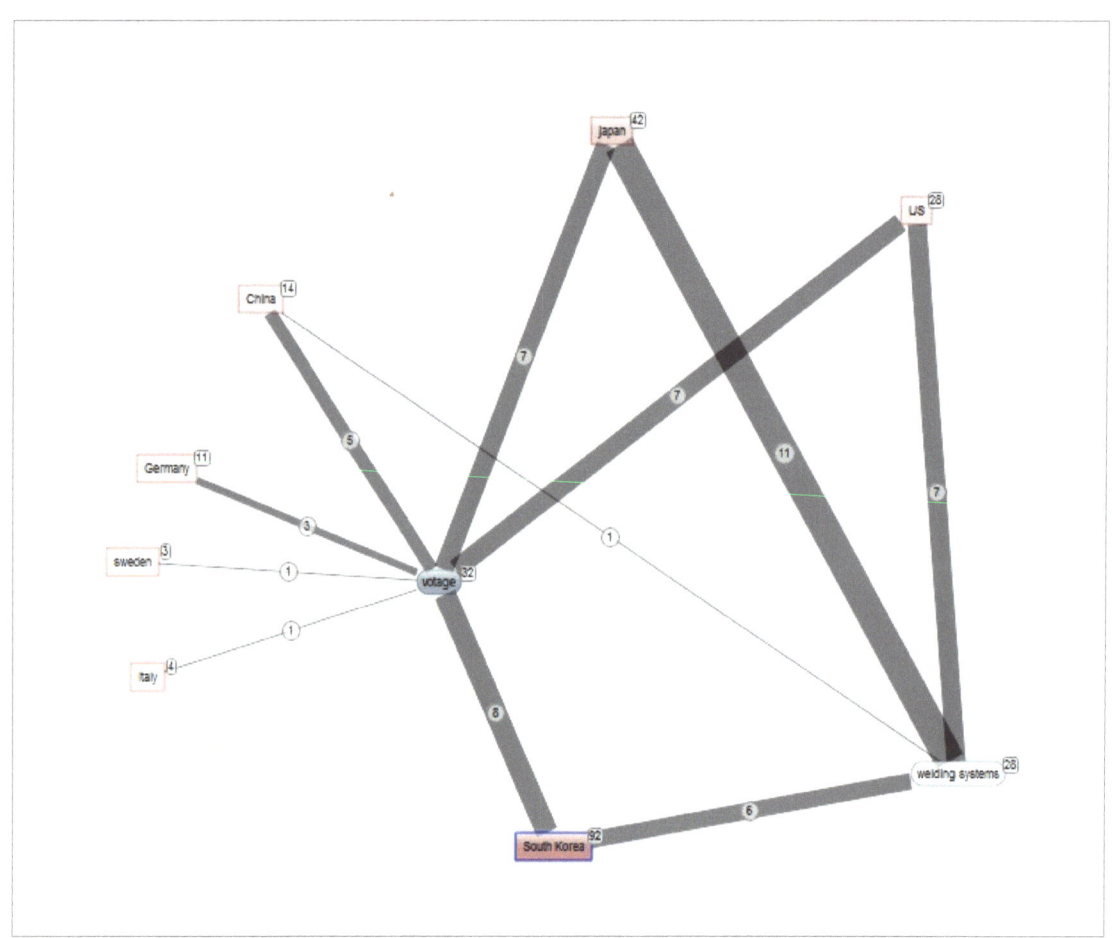

Figure 43 Involvement of the different countries in the "voltage" and 'welding systems" groups

The figure shows the difference in strategy developed by South Korea and Japan in the field of Welding Apparatus. The network of South Korea patents, Japanese patents and IPC4 digits has been draw. It shows the differences in strategy between the two countries: the links of IPC which are not shared between the two countries.

The various analysis which are presented here, show how APA (Automatic Patent Analysis) can be used to understand on a large scale what is going on in a certain area. The number of patents which is analyzed may be very large (up to several thousands) and then the analysis can be performed in details through the development of various groups according the needs of the users. In this example, we used only the results obtained for the year 2012, but if the period to analyze extends to a larger number of years the downloading can be extended. The analysis of the trends in technologies and new companies will provide more interesting results. Groups may also be built by the selection of various applicants, etc. helping to see which companies could be competing with you in the future or which may also be subject to a co-operation.

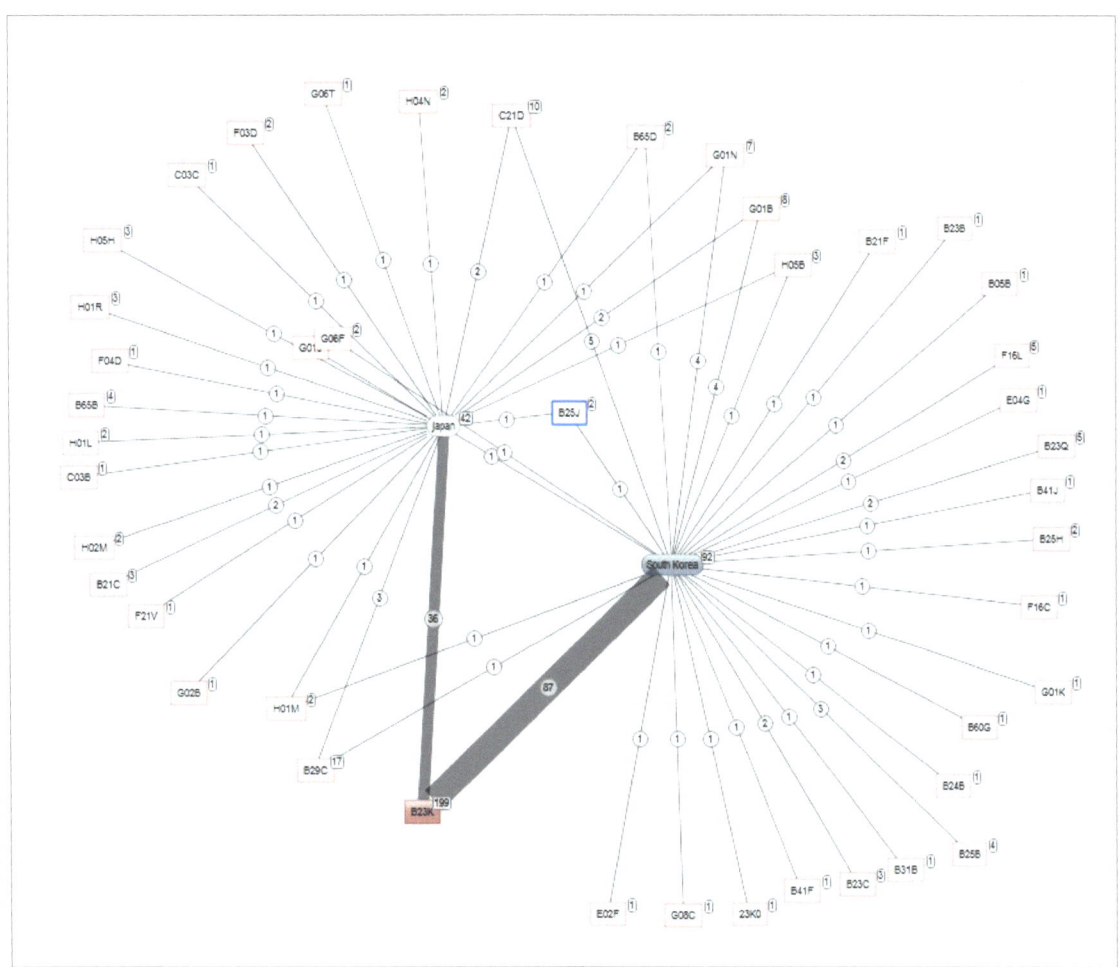

Figure 44 Comparison of strategy in R&D between Japan and South Korea

4.5 Building up reports and exporting data

Most of the time, there if a necessity to write reports and/or to exports part of the local database to intranet of co-operative networks. Facilities do exist to complete these tasks.

4.5.1 Building up reports

You have the facilities to build up automatically part of the report in different ways (reports as lists, global report, focused reports on IPC, Applicants, etc.). These reports are written automatically in Word format and you can implement these reports with your analysis by a cut and paste process for the figures and writing for the text.

Example of part of a report (format quick report):

welding [225 Families - 522 Patents]

Search #1   Administrator

Source   EPO

Easy Query: **(TI=welding and TI=apparatus) and (PD within "2012")**

331 Applicants; 556 Inventors; 83 IPC4; 310 IPC; 201 ECLA;

PD Year : 2012 [225] 2011 [34] 2010 [14] 2009 [3] 2008 [2] 2006 [1] 2005 [1] 2007 [1]

App. : empty field (--) [65] daewoo shipbuilding and marine (kr) [10] samsung heavy ind (kr) [8]

Inv. : empty field (--) [86] fujiwara junji (--) [4] kawamoto atsuhiro (jp) [4] mukai yasushi (--) [4] kowa masaru (--) [4] kawamoto atsuhiro (--) [4] fujiwara junji (jp) [4] kowa masaru (jp) [4] mukai yasushi (jp) [4] hansen richard k (us) [3]

IPC4 : b23k [199] b29c [17] empty field [12]

IPC7 : b23k9 [88] b23k37 [80] b23k26 [32]

IPC : b23k37/00 [34] b23k37/04 [30] b23k9/095 [25]

EC : empty field [164] b23k11/11b [6] b23k9/095d [6] b29c66/1122 [4]

APP/IPC : empty field (--)/b23k [58] daewoo shipbuilding and marine (kr)/b23k [10] samsung heavy ind (kr)/b23k [8] empty field (--)/b29c [7] empty field (--)/empty field [7]

Groups : voltage (--) [32] welding systems (--) [28] favorites [0]

4.5.2 Exporting data

When you have a selection of patents which is significant to you, it is possible to export these data in various formats: text, CSV, XML, for instance. Before exporting the data, you can select the fields that you want to export (including the abstract). The following example shows part of an export of a set of patents in a text format:

```
-1-
TI_EN - WELDING DEFORMATION CONTROL METHOD FOR JOINT BETWEEN LARGE INSERTED OBLIQUE TUBE AND
CYLINDER
TI_FR - PROCÉDÉ POUR RÉGULER LA DÉFORMATION AU SOUDAGE POUR UN JOINT ENTRE UN GROS TUBE OBLIQUE
EMBOÎTÉ ET UN CYLINDRE
PN - WO2012129995A1
PD - 04/10/2012
PA - SHANGHAI BOILER WORKS CO LTD (CN); FU YUWEN (CN); ZHANG BODI (CN); WANG JINING (CN)
IN - FU YUWEN (CN); ZHANG BODI (CN); WANG JINING (CN)
PR - CN20111077466 20110330


-2-
TI_EN - AIR COOLING DEVICE FOR RAPIDLY COOLING WELDING JOINT AND APPLICATION THEREOF
TI_FR - DISPOSITIF DE REFROIDISSEMENT PAR AIR POUR REFROIDISSEMENT RAPIDE DE JOINT DE SOUDURE ET
APPLICATION ASSOCIÉE
PN - WO2012129845A1
PD - 04/10/2012
PA - UNIV WUHAN SCIENCE and ENG (CN); WANG HONGHONG (CN); WU KAIMING (CN)
IN - WANG HONGHONG (CN); WU KAIMING (CN)
PR - CN20111078044 20110329


-3-
TI_EN - ON-LINE PROCESS FOR ENHANCING PERFORMANCE OF WELDING HEAT AFFECTED ZONE
TI_FR - PROCÉDÉ EN LIGNE POUR L'AMÉLIORATION DE LA PERFORMANCE D'UNE ZONE AFFECTÉE PAR LA CHALEUR DE
SOUDAGE
PN - WO2012100577A1
PD - 02/08/2012
PA - UNIV WUHAN SCIENCE and ENG (CN); NANJING DRAGON STEEL PIPE CO LTD (CN); WANG HONGHONG (CN); WU
KAIMING (CN); QIAN YONG (CN); HU FENG (CN); LEI XUANWEI (CN); TONG MINGQIANG (CN); HUANG GANG (CN)
IN - WANG HONGHONG (CN); WU KAIMING (CN); QIAN YONG (CN); HU FENG (CN); LEI XUANWEI (CN); TONG
MINGQIANG (CN); HUANG GANG (CN)
PR - CN20111025830 20110125
```

4.6 Selection from drawings

Various patents present in the first page a drawing. A database of these drawings is done automatically, and then the user instead of using various ways to select the right patents or to see what is going on in certain part of the selected patents, mays go directly to the drawings and from them, select the right patents. Groups may also be built up this way.

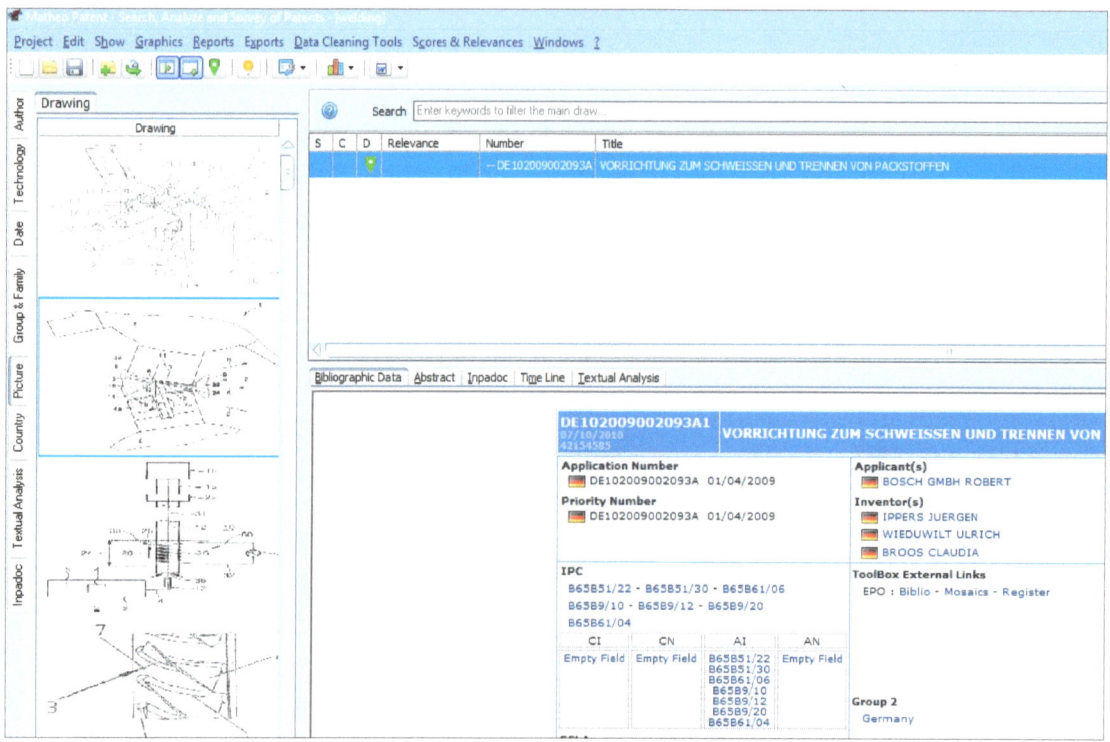

Figure 45 xample of a selection of one patent from drawings.

4.7 The cited patents

The US patent office provides for the examined patents a list of cited patents (if it is necessary during the exam of the invention) to relate it to some former patents. This is the same for the European patents (EP) and the World patents (WO). This facility provides a way to get more information on a particular area of the technology since the examiner works will provide an overview of the subject treated by this patent. There are also sometimes references to academics works which are also used. In APA drawing the cited patent network is one of the ways to get more information on a subject covers par one US or EP or WO patents if the cited patents are presents. Table 2 and figure 11 shows the cited patent network for one US patent: US201113175573A   Application number 01/7/2011

There is in table the list of the literature and patents cited:

Non-Patent Literature

- BJORKMAN ET AL. "TOOL FOR FRICTION STIR TACK WELDING OF ALUMINUM ALLOYS." NASA TECH BRIEFS
MAR. 1
VOL. 27
NO. 3
P. 65
- BROWN ET AL. "SELF-REACTING FRICTION STIR WELD FOR ALUMINUM COMPLEX CURVATURE APPLICATIONS
" AEROMAT 2003
DAYTON
JUN. 2003
31 PAGES
- BJORKMAN ET AL. "SELF-REACTING FRICTION STIR WELDING FOR ALUMINUM ALLOY CIRCUMFERENTIAL WELD APPLICATION
" ASM AEROMAT 2003
28 PAGES

Cited Document(s)

US7404512B2 - US4063676A - US5460317A - US5697511A - US5718366A - US5769306A - US5794835A
US5813592A - US5971252A - US5975406A - US6045027A - US6070784A - US6138895A - US6206268B1
US6213379B1 - US6227430B1 - US6230957B1 - US6237835B1 - US6247634B1 - US6257479B1 - US6364197B1
US6367681B1 - US6398883B1 - US6460752B1 - US6484924B1 - US6516992B1 - US6543671B2 - US6660106B1
US6742697B2 - US6848163B2 - US6913186B2 - US6955283B2 - US6986452B2 - US6994242B2 - US7156276B2
US7201811B2 - US7255258B2 - US7347351B2 - US7441686B2 - US7451661B2 - US7487901B2 - US7537150B2
US7581665B2 - US7617965B2 - US7624910B2 - US7641739B2 - US7681773B2 - US7690164B2 - US7703654B2
US2007075121A1 - US6259052B1 - US6450395B1 - US6837311B1 - US2001040179A1 - US2003217994A1 - JP2005329463A
US3704503A - USRE28709E - US5673843A - US5697544A - US5971247A - US6464127B2 - US6537682B2
US6568582B2 - US6779707B2 - US6780525B2 - US6908690B2 - US6953140B2 - US7000303B2 - US7163136B2
US7571654B2 - US7699206B2 - US7753252B2 - US7841504B2 - US7866532B1 - US7896216B2 - US3414950A
US3702913A - US3705453A - US3901497A - US3910480A - US3944202A - US3952936A - US4081651A
US4356615A - US4380348A - US4542276A - US4750662A - US5126523A - US5415435A - US5435479A
US5975405A - US6098866A - US6273320B1 - US6459062B1 - US6840433B2 - US7398909B2 - US2005120534A1
US2006289608A1 - US2010230470A1 - US6924452B2 - US7195143B2 - US7540401B2 - US7780065B2

Figure 46 Non-Patent literature and cited patents (Documents) for the US Patent US201113175573A   Application number 01/07/2011

70

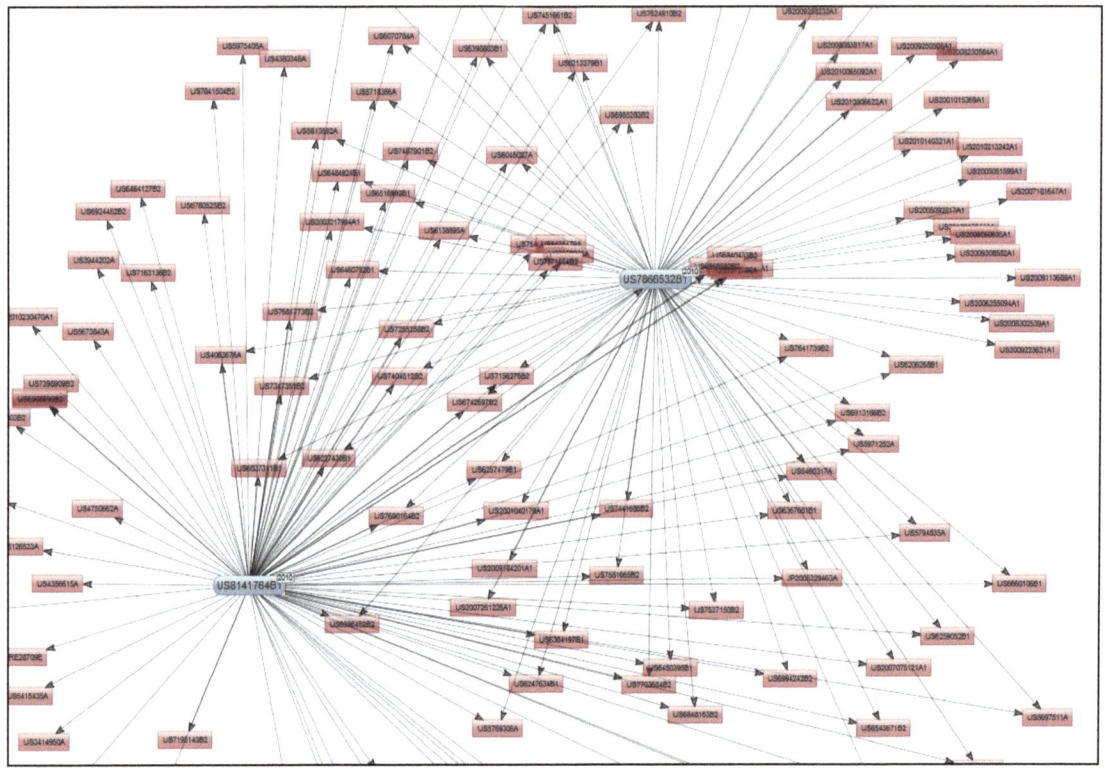

Because some cited patents are also cited patents of another patent presents in the local database the two networks are presents altogether. Various works have been done about patent citations globally two main points are underlined:

- access to more patent information related to the subject and which most of the time are not present in the original query
- More a patent is cited more important it is. This provides sometimes a to detect important patents.

Cited patents by the patents present in the local database can be downloaded and added to this database. They can be differentiated in a special group. This provides a simple way to enrich the primary local database.

8 – The strategic dependence

An invention can be protected by a patent. This patent will be the first patent granted and that will be the priority patent. Generally for a given country, most of the patent

71

with this country as priority country are patents from local enterprises, but it is possible to make the difference between both.

A patent may also (there is a delay of 12 months to do it) extended in another country. This patent will be for instance a Brazilian patent, but its priority will be from another country and then from a company which is not a Brazilian company. The APA treatment allows to differentiating these two kinds of patents. Then, by building up a matrix between the PR=BR (patents with the Brazil as priority country), the for instance US patents extended to Brazil (PN=BR but PR=US) and the IPC it will be possible to see the percent of dependence of Brazil to USA in the different domains cover by the IPC present in the patents.

4.8 Conclusion

Patents analysis provides a fast and accurate ways to understand the global picture of an area of R&D. Because the patents are evaluated by independent specialists and because the applicant pay to get its patent granted, patents represent a very valuable source of information. Today, because the main patent database are part of the open source information and because the number of patents is growing each year, the synergy between patent data and APA provides a modern way to show rapidly what are the different developments in your domain of competence. This opens the way to pre-clusterization and to public and private partnerships. The facility to get patents information will hopefully induce academics to use more patent information in their research and bibliographic references and then opening the way to closer academics and Industrial relationships.

4.9 Bibliography

Dou Henri, Léveillé Valérie, Manullang Sri, Dou Jean-Marie r, Patent Analysis for Competitive Technical Intelligence and Innovative Thinking, Data Science Journal (DSJ), Vol. 4 (2005) pp.20: a tool to improve innovative thinking, World Patent Information, Volume 26, Issue 4 , December 2004, Pages 297-309

Erikson Per, (2006), Strategic Intelligence and Innovative Clusters – A Regional Policy Blueprint Highlighting the use of Strategic Intelligence in Cluster policy. Interreg III C (European Community) Centro Formativo Privinciale, Guiseppe Zanardelli, Azienda speciale de la provincia de Brescia, Interreg III C, VINNOVA, Brics-workshop - Aalborg Swedish Governmental Agency for Innovation Systems, 13th Feb 2006

Guellec Dominique,, Bruno van Pottelsberghe de la Potterie, (2001), The internationalisation of technology analysed with patent data, n° 30, pp.1253–1266

http://gb.espacenet.com/search97cgi/s97_cgi.exe?Action=FormGen&Template=gb/EN/home.hts

http://www.matheo-software.com a trial version is available free of charge

IPC (2012), http://www.wipo.org/classifications/fulltext/new_IPC/index.htm6

Inpadoc (2012) Present the legal status of patents;
http://www.epo.org/searching/subscription/raw/product-14-11_fr.html

Navery Nicholas, (2010), Public and private Partnerships, second edition, Editor Lavoisier S.A.S.

Rostaing Hervé, (1996), La bibliométrie et ses techniques, Sciences de la Société, Collection "Outils et Méthode

WIPO (2012), http://www.wipo.int/portal/index.html.en

Présented at the INPI (France) and « Institut des Sciences de l'Information et de l'Innovation « University Pasquale Paoli, Corté, France, June 6th 2012

5 Patent information to increase innovation in SMEs and induces pre-clusterization of the stakeholders in the field of essential oils. The case of the Rosemary (rosemarinus) in Corsica.

Henri Dou (*),Emmanuelle de Gentili (**)

(*)Directeur Atelis (Strategic Workromm of the Groupe ESCEM (France Business School)), France. douhenri@yahoo.fr

(**) Institute of Information Science and Innovation (IS2I) , University Pasquale Paoli, Corté - France   m2gentili@orange.fr

Abstract: patents are for most users a tool for protecting intellectual property. But this is only a facet of their use. Indeed, the APA (Automatic Patent Analysis) allows the use of patent information (over 90 million records covering more than 90 countries) to provide strategic elements (matrices, networks, lists, etc ...) which allow to position a subject in the context of technologies and applications highlighting new applications, potential partners, etc. .. In this paper we present real a case study, which focuses on one of the Corsica potential development: the essential oils. This example will show how the right information may attract the SMEs interest as well as the one of academics, developing new links between all the stakeholders of this area. The subject chosen as an example is the rosemary (this is one of the natural plant which can be used in cosmetics and para-pharmacy. The Corsica's rosemary present because of the island weather conditions a high concentration of essential oil. We used the EPO worldpatent database to extract the data using as a tool the Matheo-Patent software (to develop the local database and to perform the APA (Automatic Patent Analysis). The time period used was from 2000 to present (May 2012). In this case study, the various applications are described, the strategic relationship between applications, applicants (industries) are examined as well as the technologies necessary for the realization of certain products. These results will provide a dashboard for Corsican's SMEs showing that with their competencies

(augmented by the local academics ones) and the local natural resources they may create new
high value products. This simple fact opens the way for a regional pre-clusterization.

Key words: Competitive Intelligence, APA (Automatic Patent Analysis), Regional Development, Bibliometrics, PPP Public and Private partnerships, Cluster, Cosmetics, Natural products

5.1   Introduction

The development of innovation most of the time is done in two steps: (Erikson 2006)

The first one is the development of competences and knowledge in Universities and Research centers mainly with the financial support of the Government.

The second one is at least also important than the first, is to transform these knowledge and competences in products and services with the help of the industry.

This classical model gives rise to clusters and in France to the poles of competitiveness where the intersection of the Research, States Institutions and Industry provides the best conditions to achieve the second step. Taking the problem that way means that to develop a cluster it is necessary to go from a scattered state of actors to a pre-clusterized one. Different works have been done about this subject, but we believe by experience, that to attract the attention and concern of the stakeholders it is necessary to create a strong incentive and to show that the sector (which can be economic, industrial, touristic ...) will benefit from this experience. It is ten necessary to find "something" which can bridge the gap between research, state institution and industry.

Patents, which open a large window on applications and products development (Yanhong 2007), (Guellec 2001), (Zoltan 1998) can be used in this case, not to protect an invention, but to provide a way to underline what others (university , industry, research centers, individuals...) develop from in our case a common resources (eg natural plants occurring in Corsica). It is also interesting to see that what is published in patent is most of the time not published elsewhere and also that patents are practically never cited in academics bibliography. Then the patent analysis dealing with a specific field prefigures a virtual cluster which, for the local stakeholders will prefigure also what can be locally developed. Various aspects of the use of text mining or APA or biblbiometrics (Rostaing 1996) provide examples of their potential aggregating power.

5.2 Material and method

5.2.1 Material

We use as an information source the database of EPO (European Patent Office) (Espacenet 2012) OPSVI which allows the query of the database by robots. The robot, here the Matheo-Patent software (Matheo 2012) allows to query the database, to build up a local formatted database and to perform automatically the APA (Automatic Patent Analysis). Various papers have already been published and they described the main functions of the above software (Dou 2005, 2004)

5.2.2 Methodology

The query of the database is done by using the words rosemary and rosemarinus (the query is made by using the classical and also the botanic names) in the patent titles (to be more precise that if the names are used also in the abstracts). This will avoid all the patents were the word rosemary or rosemarynus are cited among a lot of other aromatic plant names, without a specific application). The query has been limited to the interval from 2000 to date (June 2012). If it is necessary it is possible to extend the search (using abstracts or to extend the interval of time, for instance to retrieve all the patents with more than 20 years which are in the public domain (WIPO 2012). A local database of 162 patents is then built up. On this database the following treatments will be performed:

All analysis and correlations (lists, matrix, networks) which can be built from the documentary fields available in the downloaded patents (titles and abstracts words, Patent dates, number, priority, IPC (International Patent Classification from 4 or 8 digits) (IPC 2012)

Creation of patent groups according a technology (IPC or significant expression of titles and abstracts), dates, countries, etc. (for instance group of antioxidant, or European patents, or super-critical extraction, etc.)

Moreover on each group or on various parts of the local database the following treatments will be available:

- All analyses and correlations (with the documentary fields present)

76

- Creation for each important patent of performance index and a comment if necessary (this will facilitate the work of different experts on the same set of data)
- Export of any part of the data in text, XML or CSV formats
- Realization of various automatic reports in Word format from the local database (full, IPC, Patent Assignees, etc.)
- The update of the database at any time

Among the patents retrieved a particular attention will concern the patents which are not extended to France, this is the case for various Chinese patents (Dou 2012). This is because beyond the 12 months of possible extension, all the ideas, detail, etc. described in these patents will be available free (WIPO 2012).

This study can also be extended using the same software to the two databases of the USPTO (US Patent and Trademark Office), patent granted and patent on demand.

5.3 Functions of the Matheo-Patent software

5.3.1 General functions

The main screen after downloading presents the titles of the patents, one mouse click on the title gives access to the full title, abstracts, significant expressions from titles and abstracts, geo-mapping if available, etc. The full text of the patent can also be available using a right click of the mouse on the title. If the title if preceded by the + sign, clicking on the sign will open the patent family members. All these facilities are presented in the following figures.

The figures present most of the automatic analysis which can be performed on the local database or on various groups selected patents.

The local database can also be queried locally on all the available documentary fields (Titles, Abstracts, inventors, Patent Assignees, IPC, etc.).

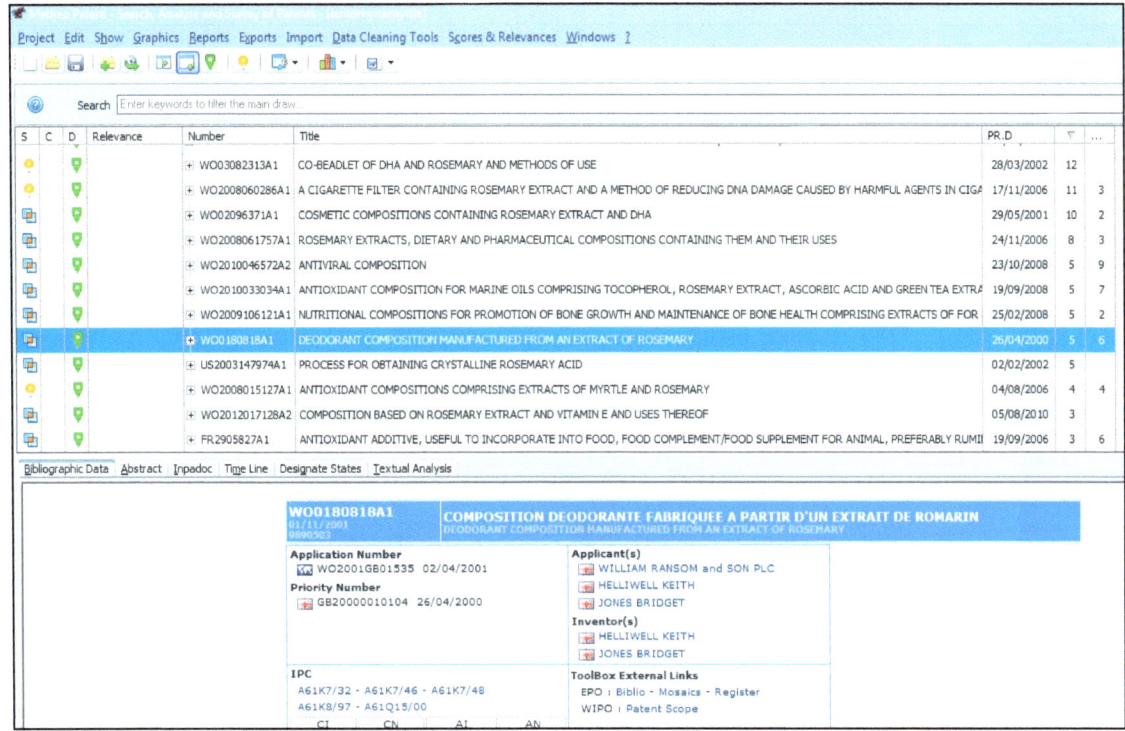

Figure 48 Main screen and access to the bibliographic information about a selected patent (in blue)

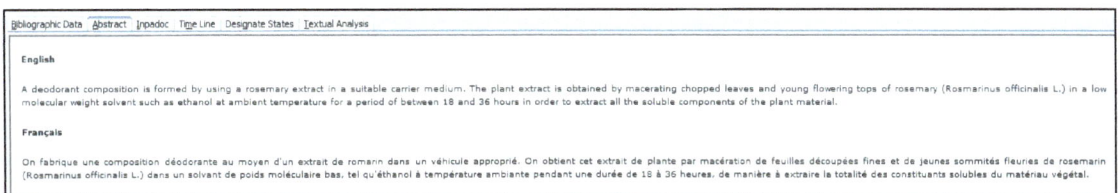

Figure 49 Main screen, access to the abstract (in several languages if available)

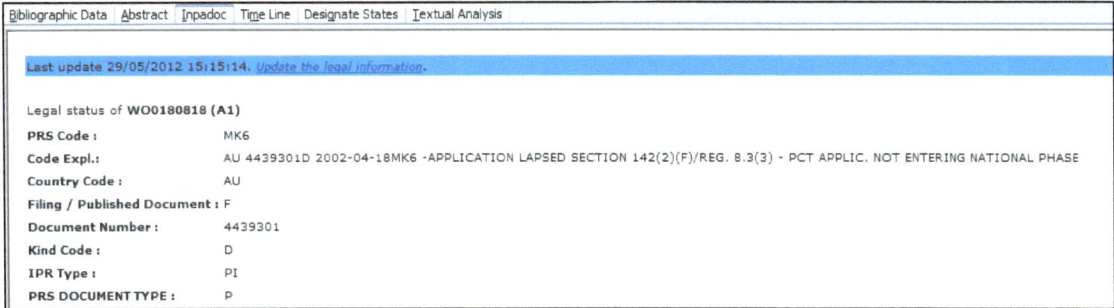

Figure 50 Main screen, Inpadoc data if available (Inpadoc 2012)

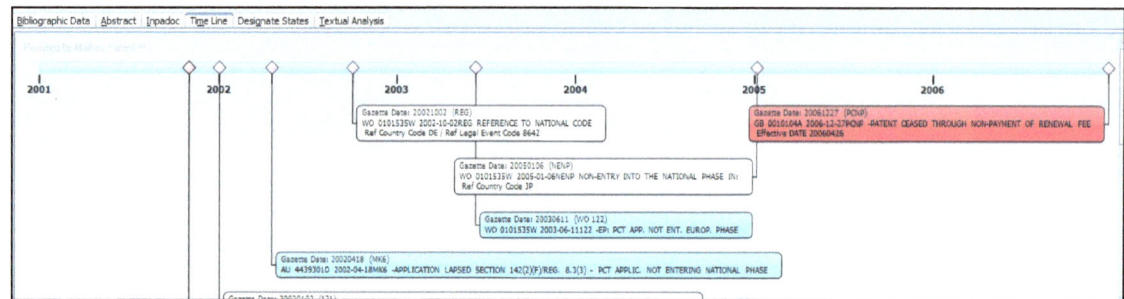

Figure 51 Main screen, access to the « life of the patent » (chronology of events)

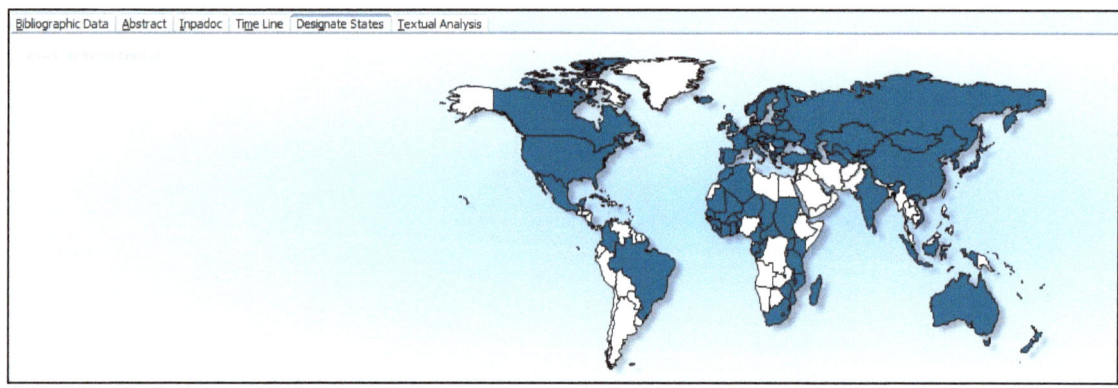

Figure 52 Main screen, mapping of the designated states

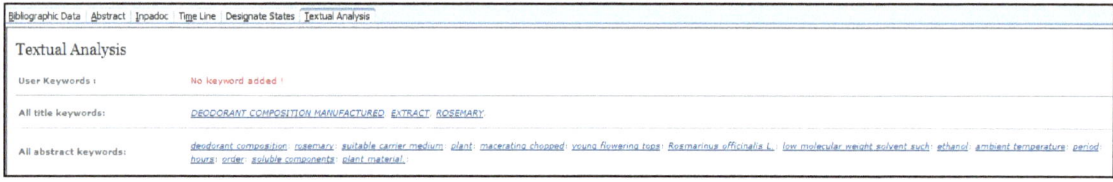

Figure 53 Main screen, significant expressions extracted from the title and the abstract

5.3.2 Examples of various analysis

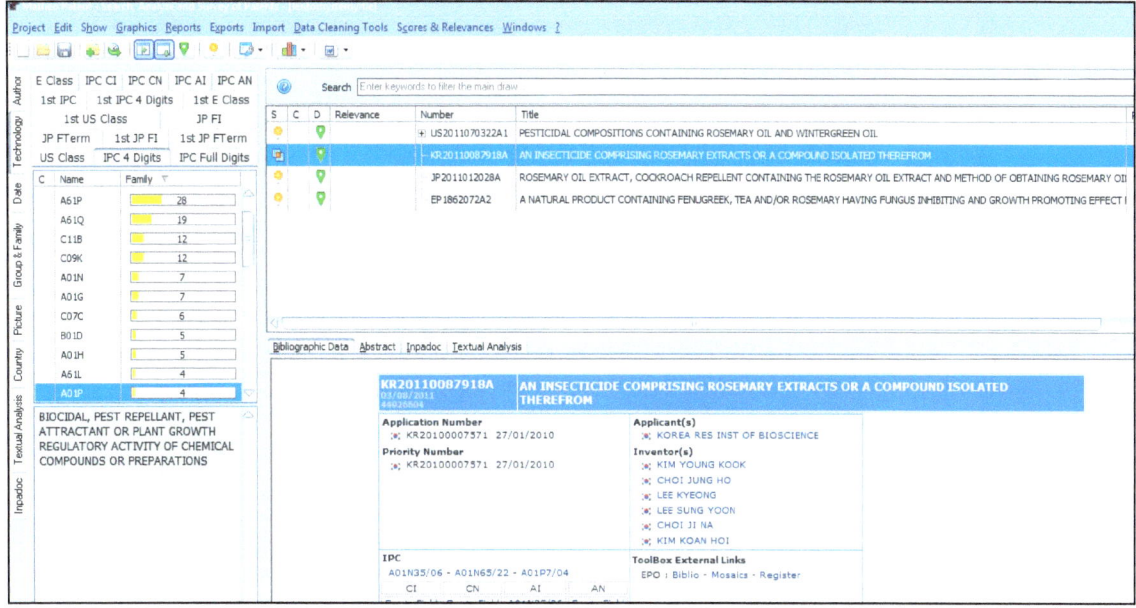

Figure 54 Analysis of the IPC 4 digits. Selection of one IPC patent lists (left) -> section on a patent

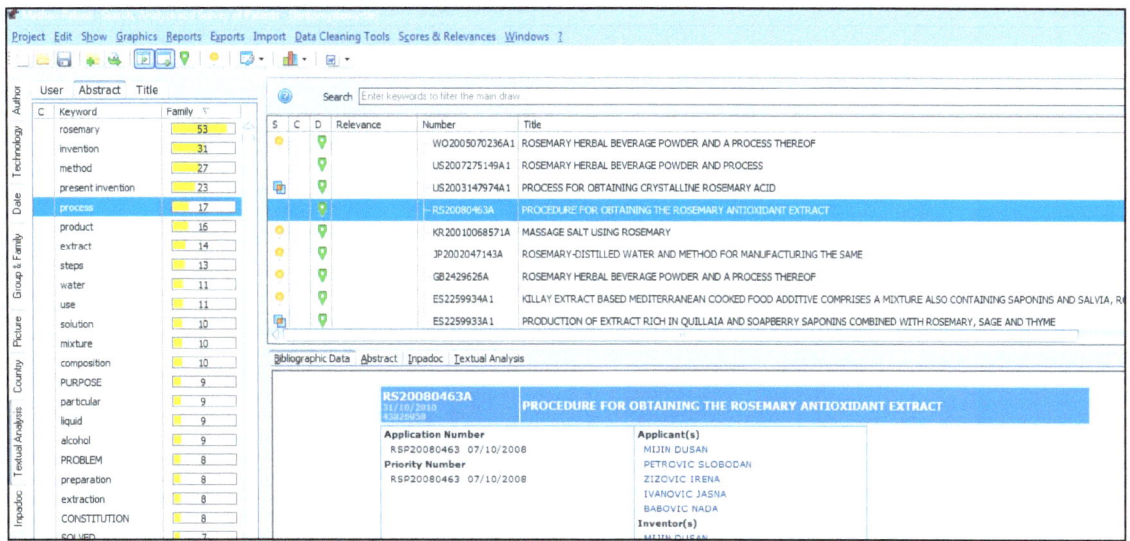

Figure 55 Textual analysis using significant expressions from titles and abstracts

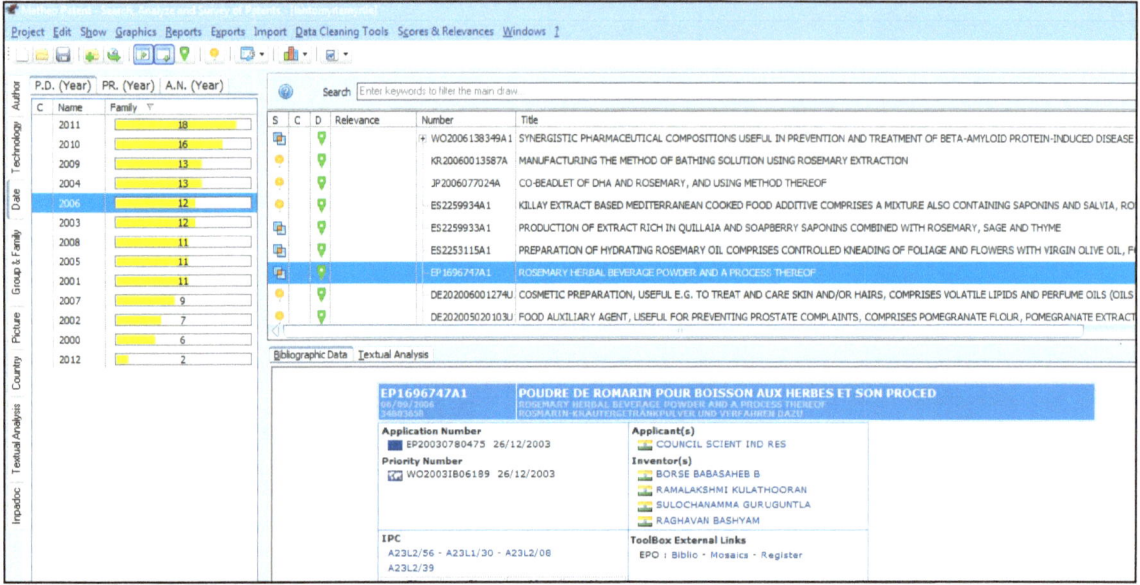

Figure 56 Chronology of the patent dates

From these various treatments (elaborated information), various groups of patents can be done. These groups will be further analyzed until the moment of the most important patents will be selected for reading. The following figures present various examples.

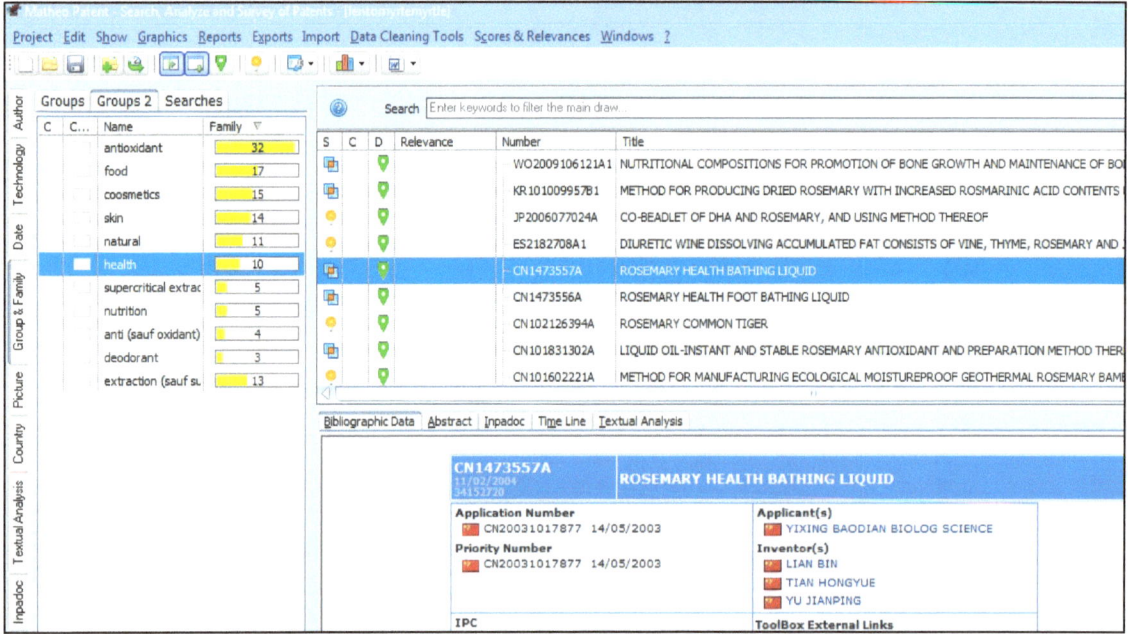

Figure 57 Examples of various groups built up from the former analysis

5.3.3 Specific analysis on the full local database

5.3.3.1 Examples

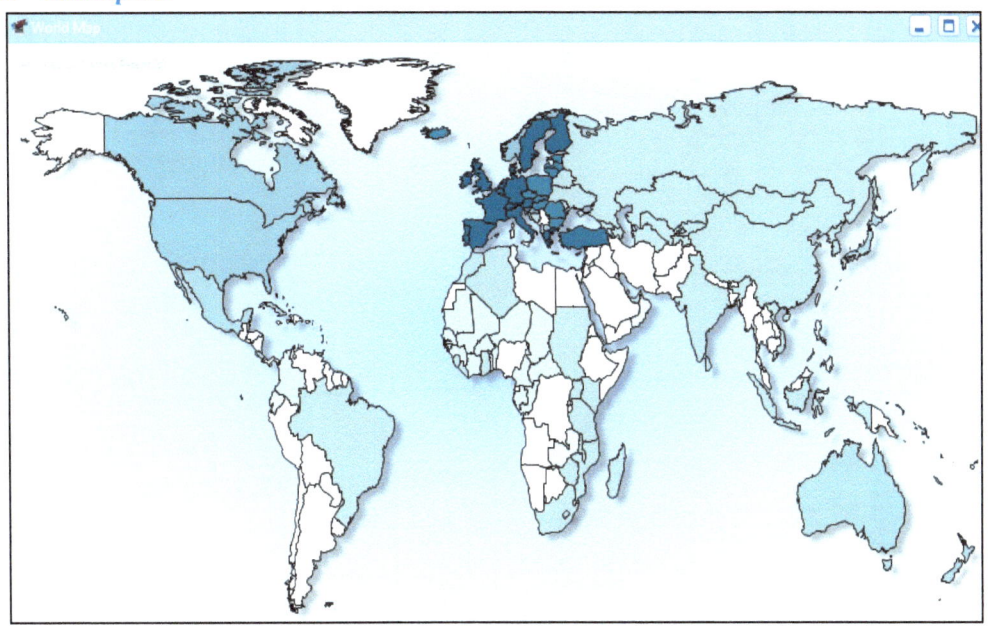

Figure 58 Designated states in the full local database

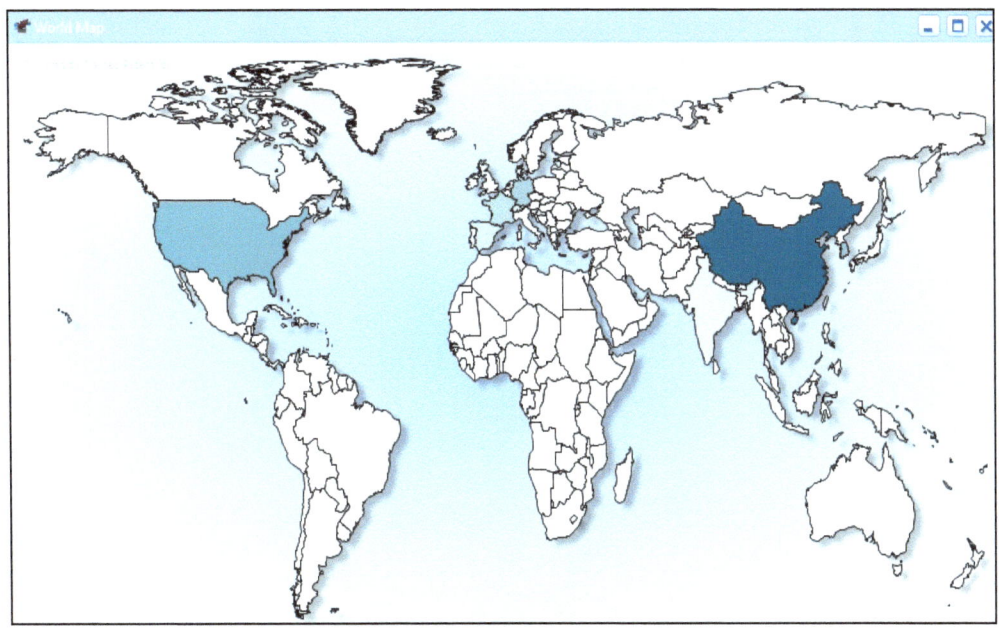

Figure 59 Patent Assignees countries present in the full local database

IP Class	ALCON INC	BIODAR LTD	COUNCIL SCIENT IND RES	BIOSYNTHEC	KEMIN IND INC	WILLIAM RANSOM AND SON...	ECOSMART TECHNOLOGIES INC	E L MANAGEMENT CORP (US)	ALCON INC (CH)	BIO DAR L TD (IL)	KIM DARRICK S H L (US)	BIOSYNTHEC (FR)	EPAX AS (NO)	MANUELA GISLER (CH)	UNIV CHINA AGRICULTURA...	ZHENG GAO (CN)	JIE KEWEI	DSM IP ASSETS BV (NL)	KEWEI JIE (CN)	JIANGSU SIMBA FLOORING...	YUEQIU GUAN	GUANGDONG FOOD INDUSTR...	BIO SYNTHEC FR	CHUNYU XING	HAINAN SUPTEK BIOTECHN...	GUANGZHOU HEBO BIOTECH...	UNIV NORTHEAST FORESTRY	YUANGANG ZU	NESTEC SA	UNIV HEFEI TECHNOLOGY	GUANGZHOU HERBS BIO SC...	HUANYU YANG	YUZHOU SENYUAN BENCAO...
A61K	2	1		1	1	1	1		2	1	1		1	1				1	1	1				1					1				3
A23L	2	1	1	1					2	1			1					1											1				
A61P	2	1							2	1			1					1	1										1				3
A24D				1								2										1											
A01G					1																												
A61Q						1		1										1						1									
A01N							1				1																1	1					
C07C											1																			1	1		
B01D				1									1									1				1							
C11B														1												1					1		
A23D														1																			
C09K														1	1	1							1			1					1		
B27M																									1								
B27D																									1								
B27N																									1								
A23F																							1										1
C07J																										1							
C10B																											1	1					

Figure 60 Automatic benchmarking of Patent Assignee know how.(full local databse)

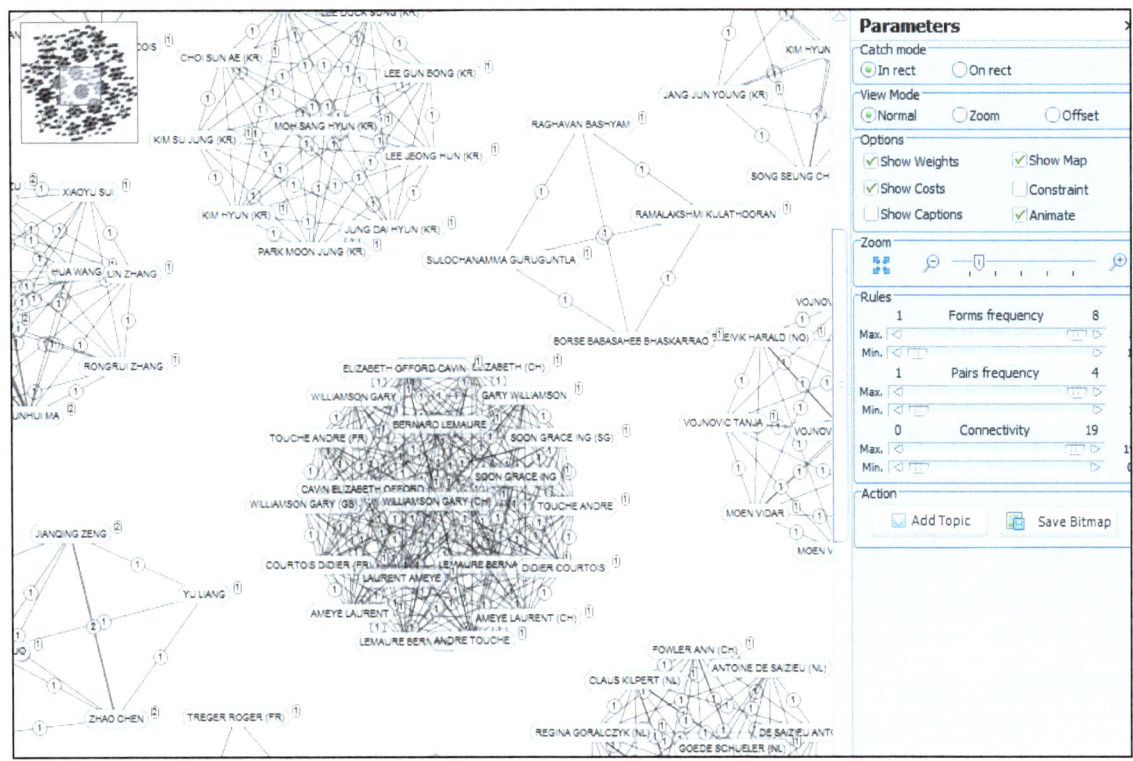

Figure 61 Partial vie of inventors networks (full local database)

5.3.3.2 Specific analysis and strategic matrices

What are the best actors in the selected domains ?

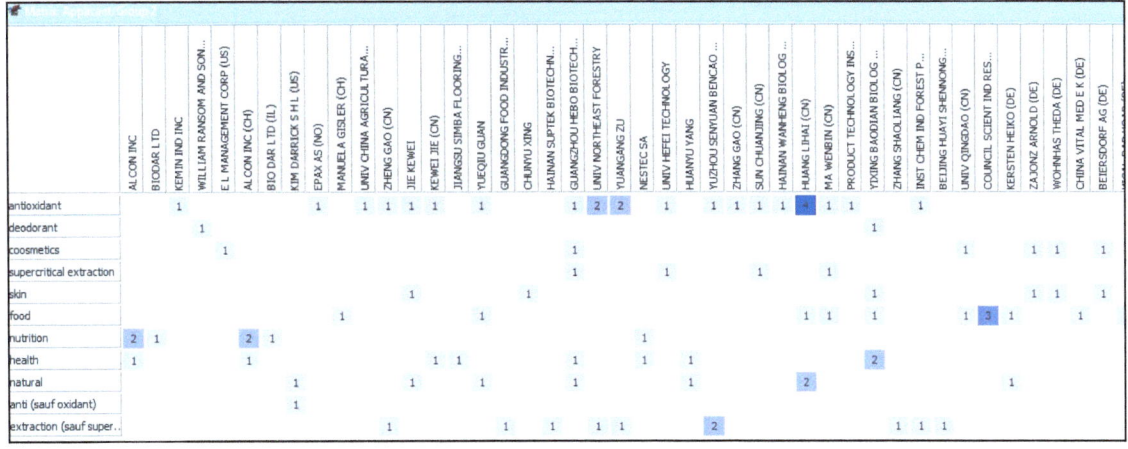

Figure 62 Partial view. The best company is the company which possess the maximum of patents in the maximum of domains

84

What is the difference in applied research between USA and China in the field of Rosemary ?

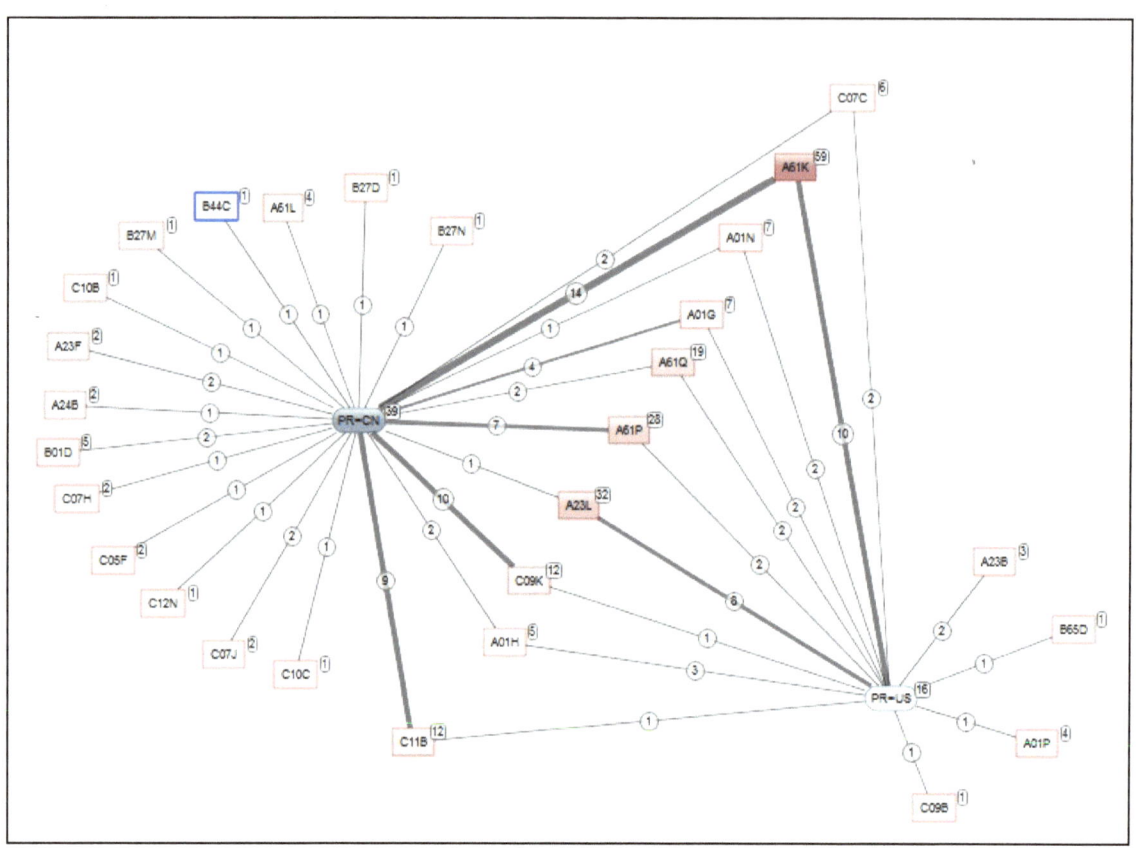

Figure 63 The specific domains are those which have only one link (UAS or China)

The same network can e done with the IPC 8 digits.

What are the application domains covered by the world (WO) patents ?

85

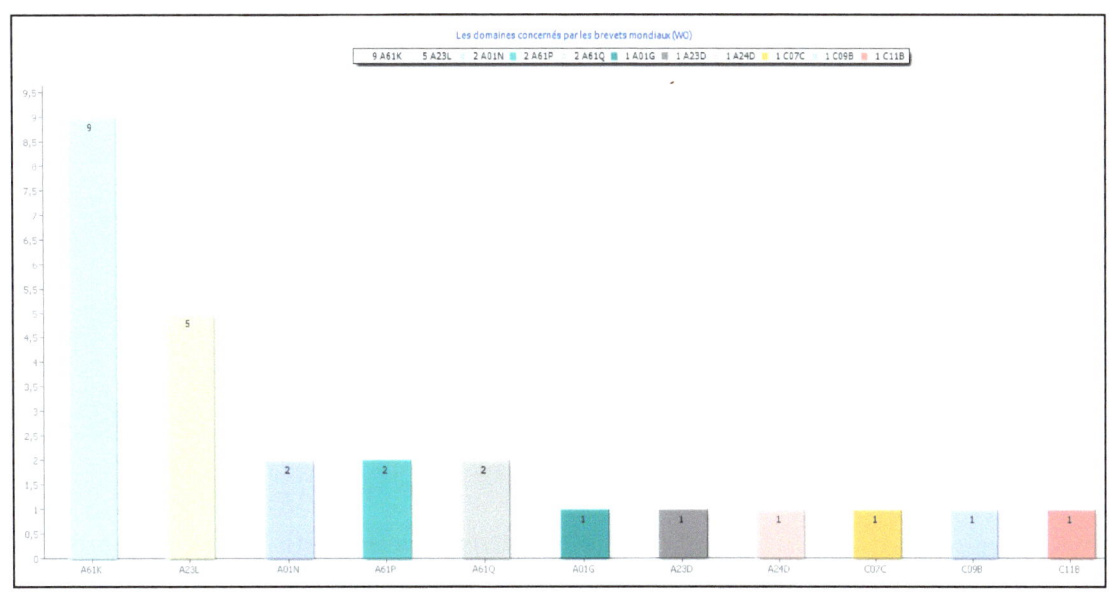

Figure 64 Technologies covered by the world patents (WO)

Priority country and strategic domains of application

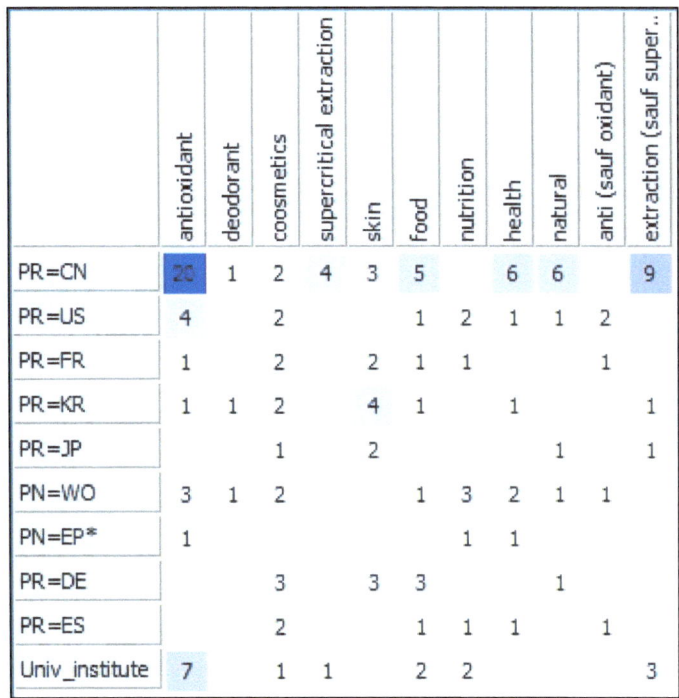

	antioxidant	deodorant	coosmetics	supercritical extraction	skin	food	nutrition	health	natural	anti (sauf oxidant)	extraction (sauf super..
PR=CN	20	1	2	4	3	5		6	6		9
PR=US	4		2			1	2	1	1	2	
PR=FR	1		2		2	1	1			1	
PR=KR	1	1	2		4	1		1			1
PR=JP			1		2				1		1
PN=WO	3	1	2			1	3	2	1	1	
PN=EP*	1						1	1			
PR=DE			3		3	3			1		
PR=ES			2			1	1	1		1	
Univ_institute	7		1	1		2	2				3

Figure 65 Priority countries and strategic domains (CN=China, etc.)

Universities and research institutes, domain of research

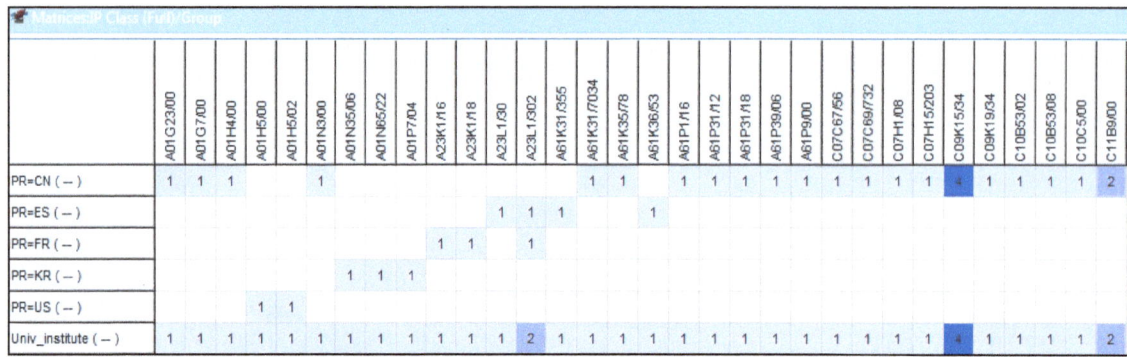

	A01G23/00	A01G7/00	A01H4/00	A01H5/00	A01H5/02	A01N3/00	A01N35/06	A01N65/22	A01P7/04	A23K1/16	A23K1/18	A23L1/30	A23L1/302	A61K31/355	A61K31/7034	A61K35/78	A61K36/53	A61P1/16	A61P31/12	A61P31/18	A61P39/06	A61P9/00	C07C67/56	C07C69/732	C07H1/08	C07H15/203	C09K15/34	C09K19/34	C10B53/02	C10B53/08	C10C5/00	C11B9/00
PR=CN (—)	1	1	1			1								1	1			1	1	1	1	1	1	1	1	1	4	1	1	1	1	2
PR=ES (—)												1	1	1			1															
PR=FR (—)										1	1		1																			
PR=KR (—)							1	1	1																							
PR=US (—)				1	1																											
Univ_institute (—)	1	1	1	1	1	1	1	1	1	1	1	1	2	1	1	1	1	1	1	1	1	1	1	1	1	1	4	1	1	1	1	2

Figure 66 Research domains of universities and institutes

This is a meta matrix. How to read it: This matrix is done by crossing the group of the priority countries (PR=xx) plus the group of the universities and institutes with the IPC 8 digits. All the applications and technologies are on top of the matrix and at the bottom the number of patents granted to institutes and universities. The rows dispatch also the patents numbers by countries of each institute and for each IPC.

Group mapping of the patents granted to universities of institutes

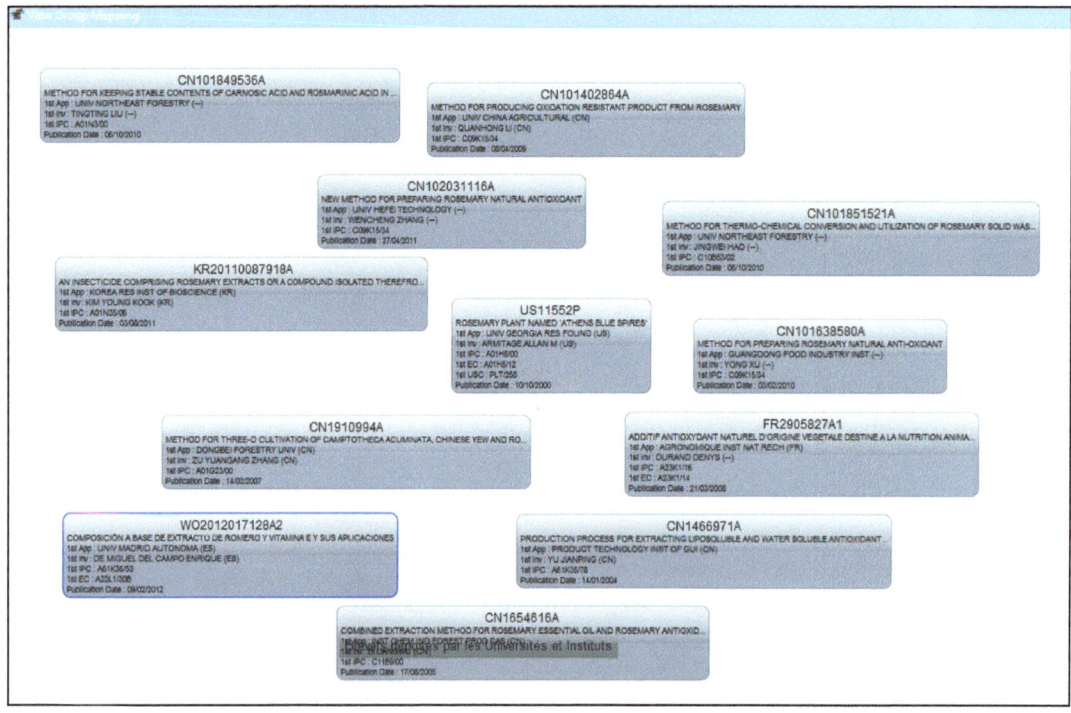

Figure 67 Patents granted to universities or institutes

87

Detailed analysis of the strategic domains (partiel view)

Figure 68 Detailed analysis of the strategic domains

Data export

The local database can globally or partially be exported in different formats (text, XML, CSV, etc .) allows for instance the integration of the data in local intranets (XML) or the export to other software (Matheo-Analyzer) to perform other statistical analysis.

Example :

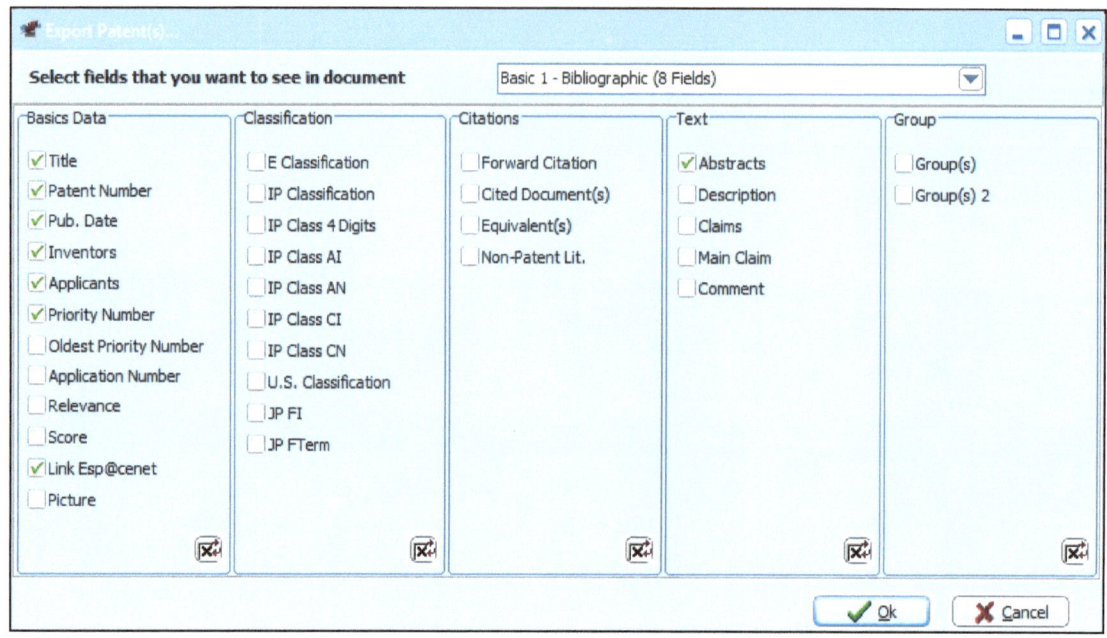

Result obtained for the strategic group: Health

-1-

TI_FR - COMPOSITIONS NUTRITIONNELLES DESTINÉES À FAVORISER LA CROISSANCE OSSEUSE ET À GARANTIR LA SANTÉ OSSEUSE QUI CONTIENT PAR EXEMPLE DES EXTRAITS DE ROMARIN OU DE CARVI

TI_EN - NUTRITIONAL COMPOSITIONS FOR PROMOTION OF BONE GROWTH AND MAINTENANCE OF BONE HEALTH COMPRISING EXTRACTS OF FOR EXAMPLE ROSEMARY OR CARAWAY

PN - WO2009106121A1

PD - 03/09/2009

PA - NESTEC SA (CH); OFFORD CAVIN ELIZABETH (CH); WILLIAMSON GARY (CH); COURTOIS DIDIER (FR); LEMAURE BERNARD (FR); TOUCHE ANDRE (FR); SOON GRACE ING (SG); AMEYE LAURENT (CH)

IN - OFFORD CAVIN ELIZABETH (CH); WILLIAMSON GARY (CH); COURTOIS DIDIER (FR); LEMAURE BERNARD (FR); TOUCHE ANDRE (FR); SOON GRACE ING (SG); AMEYE LAURENT (CH)

PR - WO2008EP52251 20080225

AB_EN - Compositions and methods for maintenance of bone health or prevention, alleviation and/or treatment of bone disorders are presented. The present invention also provides the manufacture of a nutritional product, a supplement or a medicament for promoting bone growth or for the maintenance of bone health and methods regarding same. In an embodiment, the present invention provides a composition comprising an active ingredient having an effective amount of a plant or plant extract containing at least one phytochemical having the ability to induce bone morphogenic protein expression.

AB_FR - L'invention concerne des compositions et des procédés destinés à garantir la santé osseuse ou à prévenir, soulager et/ou traiter les troubles osseux. Cette invention concerne également la fabrication d'un produit nutritionnel, d'un supplément alimentaire ou d'un médicament destinés à favoriser la croissance osseuse ou à garantir la santé osseuse, ainsi que des procédés correspondants. Dans l'un des modes de réalisation, cette invention concerne une composition contenant un principe actif possédant une quantité efficace d'une plante ou d'un extrait végétal qui contient au moins un phytochimique permettant d'induire l'expression d'une protéine morphogène osseuse.

UR -
http://worldwide.espacenet.com/publicationDetails/biblio?CC=WO&NR=2009106121&KC=A1&FT=E&DB=EPODOC&locale=en_EP

-2-

TI_EN - METHOD FOR PRODUCING DRIED ROSEMARY WITH INCREASED ROSMARINIC ACID CONTENTS USING FAR-INFRARED IRRADIATION

PN - KR101009957B1

PD - 20/01/2011

PA - KNU INDUSTRY COOPERATION FOUND (KR)

IN - CHO DONG HA (KR); JIN CHENG WU (CN); SONG SEUNG CHUL (KR); CHEI SUNG CHUN (KR); KIM HYUN SAM (KR); JANG JUN YOUNG (KR); KWON HYUK MIN (KR)

PR - KR20100043273 20100510

AB_EN - PURPOSE: A cultivation method of rosemary is provided to use the rosemary as health supplementary functional food by increasing the content of rosmarinic acid. CONSTITUTION: A cultivation method of rosemary comprises the following steps: drying young leaves and stems of the rosemary by irradiating far-infrared rays with the wavelength of 2~14micrometers in

100~120deg C for 5~20 minutes. The drying efficiency of the rosemary is enhanced during the process.

UR -
http://worldwide.espacenet.com/publicationDetails/biblio?CC=KR&NR=101009957&KC=B1&FT=E&DB=EPODOC&locale=en_EP

.....

Note the presence o the URL which allows automatically to enter in the EPO database to access information (bibliographic description, first page, drawings, full texte of the patent).

Data Import

People working with : Delphion (format XML), PatBase (format XML), Micro Patent (format CSV), Derwent (format ISI) to benefit of the Matheo-Patent treatments, Matheo-Patent integrates automatically the format the former patent management software.

5.4 Conclusion

All he former results underline how from a precise subject, he APA may show to experts and stakeholders of this domain a deep insight on what is going on in applied sciences. Because information provided by patents is unique, this information can be used to attract the different stakeholders and to prefigure a cluster development. This method has been applied in Corsica to other natural aromatic plants such as the Chrysantemun italicum (everlasting flower), the thyme, etc. The various SMEs engaged in cosmetics were very interested specially when they show that some properties for instance the cicatrization for the ever lasting flowers can be used in para-medical products. In June an exhibition of the former SMEs was done in Ajaccio, an one of the conclusion of the various meetings was to begin to develop a cluster approach for all the industries involved in the subject. For instance we extend now the study of the patents and scientific literature to the various processes of extraction to provide the best yield and purity of the essential oils. With the help of the regional parliament (area of industrial development and innovation), a grant will be provided to the Corsica University to develop next university year a global approach upon this subject by collecting the right information, (including patents) and with the help of experts to transform this information in actionable knowledge which will be integrated into a strategic project of pre-clusterization in the domain of aromatic plants, essential oils and cosmetics.

The use of academic, patent and economic information fills the gap between academic research and industrial concerns and open the way for the development of PPP Public and Private Partnerships (Academics, industry and political institutions) (Navery 2010).

5.5 Biblography

Dou Henri, (2012) Chinese Strategy in Intellectual Property Management, VSST 2012 Ajaccio, opening conference (available on http://www.ciworldwide.org)

Dou Henri, Léveillé Valérie, Manullang Sri, Dou Jean-Marie r, Patent Analysis for Competitive Technical Intelligence and Innovative Thinking, Data Science Journal (DSJ), Vol. 4 (2005) pp.209-236

Dou Henri , Benchmarking R&D and companies through patent analysis using free databases and special software: a tool to improve innovative thinking, World Patent Information, Volume 26, Issue 4 , December 2004, Pages 297-309

Erikson Per, (2006), Strategic Intelligence and Innovative Clusters – A Regional Policy Blueprint Highlighting the use of Strategic Intelligence in Cluster policy. Interreg III C (European Community) Centro Formativo Privinciale, Guiseppe Zanardelli, Azienda speciale de la provincia de Brescia, Interreg III C, VINNOVA, Brics-workshop - Aalborg Swedish Governmental Agency for Innovation Systems, 13th Feb 2006

Guellec Dominique,, Bruno van Pottelsberghe de la Potterie, (2001), The internationalisation of technology analysed with patent data, n° 30, pp.1253–1266

http://gb.espacenet.com/search97cgi/s97_cgi.exe?Action=FormGen&Template=gb/EN/home.hts

http://www.matheo-software.com a trial version is available free of charge

IPC (2012), http://www.wipo.org/classifications/fulltext/new_IPC/index.htm6

Inpadoc (2012) Present the legal status of patents;
http://www.epo.org/searching/subscription/raw/product-14-11_fr.html

Navery Nicholas, (2010), Public and private Partnerships, second edition, Editor Lavoisier S.A.S.

Rostaing Hervé, (1996), La bibliométrie et ses techniques, Sciences de la Société, Collection "Outils et Méthode

WIPO (2012), http://www.wipo.int/portal/index.html.en

Yanhong Liang , Tan Runhua,(2007) A test mining-based Patent Analysis in product innovation process, in Trend in Computer aided Innovation, ed. Noël Lean-Riva, Spinger IFIP, p. 89

Zoltan J and David B Andrestch,(1998), Innovation in large and small firms. An empirical study, The American Economic Review, vol.78, n°4, pp. 678-690

6 Patent Information - Useful Internet Addresses

EPO Espace Net http://www.epo.org/searching/free/espacenet.html

WIPO (World International Property Organization
http://www.wipo.int/portal/en/index.html

USPTO (US Patent and Trademark Office)
http://www.uspto.gov/patents/process/search/

SIPO (National Chinese Patent Office)
http://english.sipo.gov.cn/

Espace Net (EPO) Advanced Search
http://worldwide.espacenet.com/advancedSearch?locale=en_EP

Espace Net (EPO) International Classification Search
http://worldwide.espacenet.com/classification?locale=en_EP

Official Catchword Index (WIPO)
http://worldwide.espacenet.com/classification?locale=en_EP

Google Patent Advanced Search
http://www.google.com/advanced_patent_search

CIWORLDWIDE Competitive Intelligence Communauty
http://www.ciworldwide.org

INPI Instituto Nacional da Propriedade Intellectual (Brazil)
http://www.inpi.gov.br/portal/artigo/busca_patentes

Japanese Patent Search via IDPL (Industrial Property Digital Library of Japan)
http://www19.ipdl.inpit.go.jp/PA1/cgi-bin/PA1INIT

Google Translate
https://translate.google.com/?hl=en&tab=wT

Dialog Online Documentation
http://library.dialog.com/bluesheets/

7   About the author

Henri Dou is Engineer IPSOI (Petroleochemical Institute) and made his doctorate in the field of Chemistry. After a career at the CNRS as Research Director, he joined the University of Aix Marseilles III as Professor in Information Science and he developed the first French cursus in Technology Watch as well as various diplomas DU, DESS, DEA, Master, PhD in the field of Competitive Technical Intelligence, Regional Development and Bibliometrics. He participated to the developed of the Competitive Intelligence in Brazil. Among the different functions the following are among the most important : General Secretary of ChIN (Chemical Information Network UNESCO), Scientific Secretary near Bernard Gregory (General Manager of the CNRS) for the cooperation CNRS-MIT (Science and Decision), "Chargé de mission" near the Manager of the Chemical sector of the CNRS for the co-operation with Rhône-Poulenc, French representative at the

International Oceanographic Commission and later in charge of the analysis of the US coal plan development. He was Director of ATELIS (Strategic Intelligence Workroom of ESCEM-France Business School), and is French expert of the Franco-Chinese Association of Competitive Intelligence, President of the French Society of Applied Bibliometrics and Member of various editorial boards of Scientific journals. He is specialized in Competitive Intelligence, APA (Automatic Patent Analysis), Regional Development and SRR (Social Research Responsibility). He currently takes part in various activities in Indonesia, China, Malaysia, Brazil, Africa and Mexico. For more information consult: http://www.amazon.fr/Henri-Dou/e/B00AWD21WU

www.ingramcontent.com/pod-product-compliance
Lightning Source LLC
Chambersburg PA
CBHW050728180526
45159CB00003B/1161